Reviews of Physiology, Biochemistry, and Pharmacology 144

Springer-Verlag Berlin Heidelberg GmbH

Reviews of

144 Physiology Biochemistry and Pharmacology

Editors

S. G. Amara, Portland   E. Bamberg, Frankfurt
M. P. Blaustein, Baltimore   H. Grunicke, Innsbruck
R. Jahn, Göttingen   W. J. Lederer, Baltimore
A. Miyajima, Tokyo   H. Murer, Zürich
N. Pfanner, Freiburg   G. Schultz, Berlin
M. Schweiger, Berlin

With 35 Figures and 7 Tables

 Springer

ISSN 0303-4240
ISBN 978-3-662-31007-6   ISBN 978-3-540-45534-9 (eBook)
DOI 10.1007/978-3-540-45534-9
Library of Congress-Catalog-Card Number 74-3674

http://www.springer.de

© Springer -Verlag Berlin Heidelberg 2002
Originally published by Springer-Verlag Berlin Heidelberg New York in 2002.
Softcover reprint of the hardcover 1st edition 2002

Production: PRO EDIT GmbH, D-69126 Heidelberg
Printed on acid-free paper   – SPIN: 10831932   27/3130göh-5 4 3 2 1 0

Contents

Indexed in Current Contents

Contents

Phospholipase D – Structure, Regulation and Function

J. H. Exton

Howard Hughes Medical Institute and Department of Molecular Physiology
and Biophysics, Vanderbilt University School of Medicine, Nashville, TN 37232, USA

Contents

Reviews of Physiology, Biochemistry,
and Pharmacology, Vol. 144
© Springer-Verlag Berlin Heidelberg 2002

1
Introduction

Phospholipase D (PLD) is an enzyme that is widely distributed in bacteria, protozoa, fungi, plants and animals. Its principal substrate is phosphatidylcholine (PC) which it hydrolyzes to phosphatidic acid (PA) and free choline. PA can be further metabolized to diacylglycerol (DAG) by phosphatidate phosphohydrolase (PAP, also called lipid phosphate phosphatase), and to lysophosphatidic acid (LPA) by phospholipase A_2 (PLA$_2$). Phospholipase D also carries out a transphosphatidylation reaction, which is unique to this enzyme. This involves the transfer of the phosphatidyl group from the phospholipid substrate to a primary alcohol to yield a phosphatidylalcohol. This reaction occurs to a very much less extent with secondary or tertiary alcohols, giving a measure of specificity.

Many PLD isoforms have been cloned from bacterial, plant, yeast and mammalian sources. All exhibit three or four highly conserved sequences. Within two of these sequences is found the highly conserved HXKX$_4$DX$_6$GSXN motif, designated HKD (Ponting & Kerr 1996; Koonin 1996). This motif is found not only in the PLD isoforms, but in a PLD superfamily comprising bacterial phosphatidylserine and cardiolipin synthases, bacterial endonucleases, pox virus envelope proteins and a murine toxin from *Yersinia* (Ponting & Kerr 1996; Koonin 1996). Mutagenesis studies have shown that the two HKD motifs in PLD are required for catalysis and associate to form a catalytic center (Sung et al. 1997; Xie et al. 1998). Catalysis occurs by a two-step mechanism and involves the formation of a phosphohistidine intermediate (Waite 1999).

Some plant and most animal PLDs require phosphatidylinositol 4,5-bisphosphate (PIP$_2$) for activity (Qin et al. 1997; Frohman et al. 1999), and mammalian PLDs are regulated by a wide variety of growth factors, cytokines, neurotransmitters and other agonists acting through G protein-coupled receptors (GPCRs) (Exton 1999). Mammalian PLD isozymes are splice variants of two major isoforms (PLD1 and PLD2) which have M_rs of 120K and 100K, respectively. They show approximately 50% amino acid sequence identity and are widely distributed (Hammond et al. 1997; Colley et al. 1997; Exton, 1998; Meier et al. 1999). PLD1 is strongly activated in vitro by conventional PKC isozymes and members of the ADP-ribosylation factor (ARF) and Rho families of small G proteins (Hammond et al. 1997; Min et al. 1998b). In contrast, PLD2 shows weak or absent responses to PKC, ARF and Rho in vitro (Colley et al. 1997; Kodaki & Yamashita 1997; Lopez et al. 1998; Slaaby et al. 2000). Mammalian cells also contain an oleate-stimulated PLD activity (Massenburg et al. 1994; Banno et al. 1997; Lee et al. 1998), but its nature is unresolved. One form is unaffected by ARF and PIP$_2$ (Massenburg et al. 1994; Lee et al. 1998) whereas another form found in nuclei is stimulated by ARF (Banno et al. 1997). Oleate and other unsaturated fatty acids can stimulate PLD2 and synergize with PIP$_2$ (Kim et al. 1999b). Three PLD isozymes that specifically act on phosphoinositides have also been discovered (Ching et al. 1999). Two of these preferentially act on phosphatidylinositol 3,4,5-trisphosphate (PIP$_3$), while the other acts on PIP$_3$ and phosphatidylinositol 3-phosphate. None of the isozymes acts on PIP$_2$, but this phospholipid stimulates their activity. Another PLD isozyme has been reported in neutrophils (Horn et al. 2001). This has an apparent M$_r$ of 90K and is not recognized by antibodies to the PLD1 and PLD2 isozymes, but does interact with an antibody raised to a conserved sequence in these enzymes. It is stimulated by PIP$_2$ and inhibited by oleate and divalent cations.

The cellular functions of PLD remain unclear. The enzyme has been implicated in mitogenesis, endocytosis, actin cytoskeleton rearrangements, glucose transport, secretion and superoxide production. It has also been proposed to play an important role in trafficking through the Golgi. Its cellular functions relate mainly to the generation of PA, which affects the activity of many enzymes and other cellular proteins. DAG derived from PA may also activate certain isoforms of protein kinase C (PKC). LPA is a major extracellular messenger that interacts with certain types of EDG receptors. PLD is likely involved in its generation via PA formation and subsequent PLA2 action.

This review is focussed on mammalian PLDs. For consideration of plant and yeast PLDs, the reader is refered to recent review articles (Pappan & Wang 1999; Rudge & Engebrecht 1999; Munnik 2001).

2
Structure of Phospholipase D

The crystal structure of PLD from *Streptomyces* has been determined and also that of the bacterial endonuclease Nuc, a member of the PLD superfamily. Nuc is a protein of small molecular mass (16 kDa) that has only one HKD motif. It crystallizes as dimer and each monomer consists of an 8-stranded β-sheet flanked by 5 α-helices (Stuckey & Dixon, 1999; Fig. 1). Each conserved HKD sequence is positioned on two loops and, in the dimer, they lie adjacent to each other to produce a single active site. The active site residues are held in place by a network of hydrogen bonds within each monomer and between both monomers. These involve not only the conserved His, Ser and Asn, but also adjacent Glu residues. Using tungstate as a phosphate surrogate to explore residues involved in substrate binding, it was found

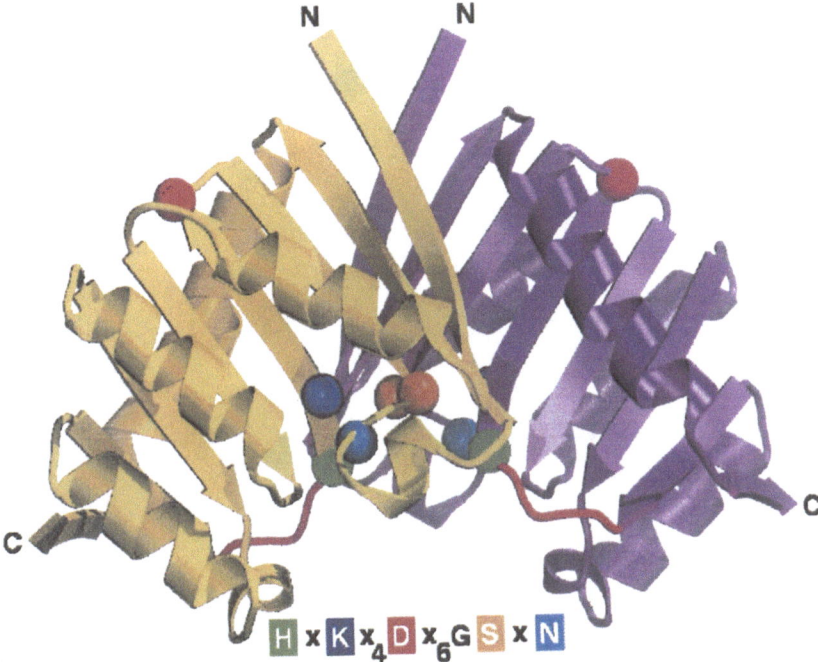

Fig. 1. Domain organization of Nuc dimer. Nuc is a bacterial endonuclease that is a member of the PLD superfamily. One monomer is yellow and the other purple with the variable loop in red. The colored spheres represent residues in the HKD motif, which are identified at the bottom of the figure. From Stuckey and Dixon (1999) by permission of the authors and publisher

that the His, The Lys and Asn residues from each HKD motif were important (Fig. 2, Stuckey & Dixon, 1999). The Lys and Asn residues are postulated to be involved in binding and neutralizing the negative charge on the phosphate of the substrate and the His residues during catalysis (see below).

The structure of *Streptomyces* PLD is shown in Fig. 3 (Leiros et al. 2000). It consists of two highly interacting components of similar topology. Each component is made up of a β-sheet composed of 8 or 9 β-strands and flanked by 9 α-helices. Overall, the structure is very similar to the Nuc endonuclease dimer. Although most of the α-helices are oriented along the β-strands, some are rotated with respect to this overall direction (Leiros et al. 2000). The active center region is very similar to that in the Nuc dimer, with identical residues from each component. The rotated α-helices are distant from the active center, but probably act as scaffolding to keep the active site open and accessible to substrate (Fig. 3, Leiros et al. 2000).

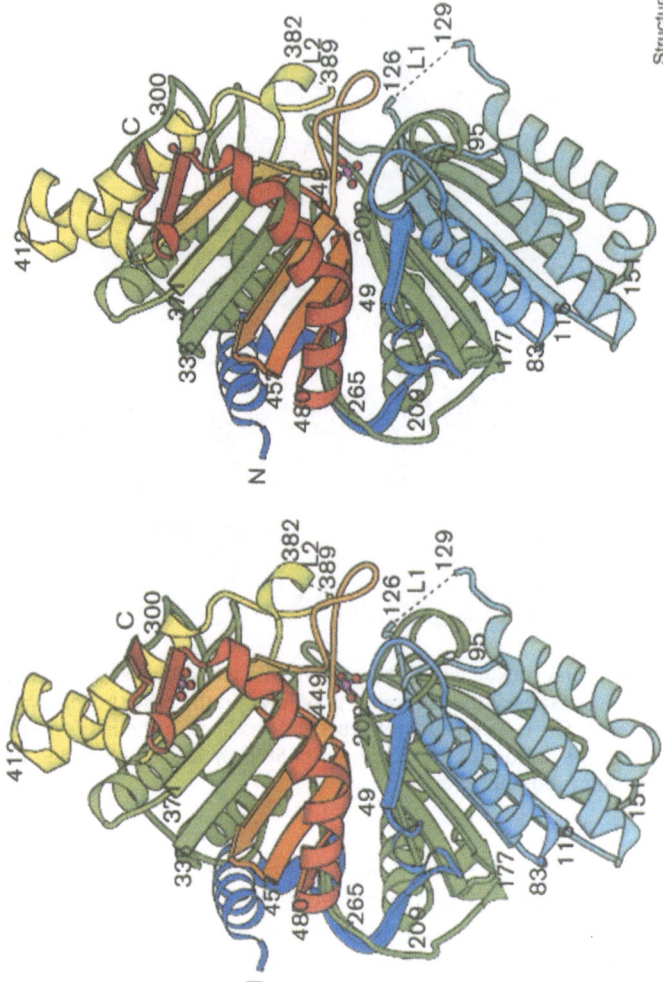

Structure

Fig. 2. Stereographic view of PLD f rom *Streptomyces* colored dark blue to red according to the sequence. Two phosphate groups are shown in red, one of which is in the catalytic center. From Leiros et al. (2000) by permission of the authors and publisher

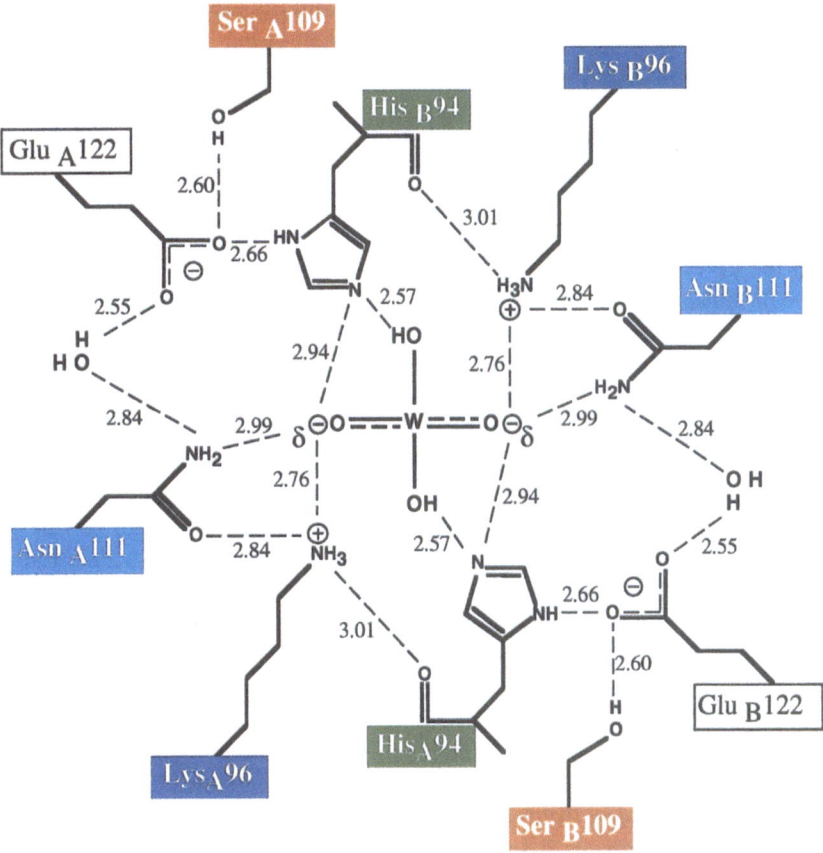

Fig. 3. Hydrogen bonding at the active center of tungstate-substituted Nuc dimer. The color coding of the amino acid residues of the HKD motifs is as described in Fig.1. The A and B subscripts refer to the different monomers. From Stuckey & Dixon (1999) by permission of the authors and publisher

The enzyme was crystallized using a citrate-phosphate buffer and one phosphate was bound at the active site (Leiros et al. 2000). This very probably corresponds to the phosphate head group of the phospholipid substrate and lies in close contact to His, Lys and Asn residues contributed from both domains of the enzyme as seen for the catalytic center of the Nuc dimer (Fig. 2). As discussed in more detail below, there is much evidence that the two His residues participate in the catalytic reaction. Interestingly, there are no structural similarities between PLD and other lipases, including the secretory and cytosolic forms of PLA$_2$, bacterial or mammalian PI phospholipase

C (PLC), and bacterial PC-phospholipase C, with the exception of a cutinase from *Fusarium* (Leiros et al. 2000). As is evident in Fig. 3, the similar topology of the two components of *Streptomyces* PLD suggests that they have a common origin. This is also suggested by the duplicated HKD motifs of the enzyme and the fact that it is possible to align the structure of the Nuc dimer against the *Streptomyces* PLD structure (Leiros et al. 2000).

The structural elements responsible for the substrate specificity of PLD remain unknown. By analogy with the structure of cytosolic PLA_2, residues on the lip of the entrance to the active site might play a role (Leiros et al. 2000). These may be located in two flexible loops located relatively close to the active center in a region thought to make membrane contact when catalysis occurs. *Streptomyces* PLD is much smaller (54 kDa) than mammalian PLDs and lacks the N- and C-termini that contain the interaction sites for PKC, Rho and ARF. Consequently, although its structure has provided very valuable information about the catalytic mechanism, detailed information about the mechanisms by which mammalian PLD is regulated by PKC and small G proteins, will await the determination of the structure of one of these isozymes. Nevertheless, mutagenesis studies have provided much information about the mechanisms of catalysis and regulation of the mammalian PLD1 and PLD2 isozymes.

Both PLD1 and 2 show four highly conserved sequences and other motifs. A schematic representation of PLD1 is shown in Fig. 4 (Sung et al. 1999b; Frohman et al. 1999). At the N-terminus are a PX domain, which is usually involved in interactions with proteins or 3-phosphoinositides, and a pleckstrin homology (PH) domain, which is frequently involved in membrane association involving PIP_2 or PIP_3. In the second and fourth conserved domains are the N-terminal and C-terminal HKD motifs. In the center of the linear sequence is a loop region, which is absent in PLD2 and is partly missing in splice variants of PLD1. A PIP_2 binding site has been identified between the loop region and the third conserved sequence (Sciorra et al. 1999). Interaction sites for PKC and Rho have been identified in the N- and C-terminal sequences, respectively (see below).

Mutagenesis studies have shown that both HKD motifs are required for catalysis (Sung et al. 1997) and co-expression studies in COS7 cells have shown that hydrophobic domains including and surrounding the motifs associate to bring them together to form a catalytic center (Xie et al. 1998, 2000a). Conserved sequence III is required for catalysis, but its role is unknown. Because of its enrichment in aromatic residues, which are also found for the choline-binding region of the acetylcholine receptor, it has been suggested that this sequence interacts with the choline headgroup of PC (Sung et al. 1997, 1999b).

human PLD1

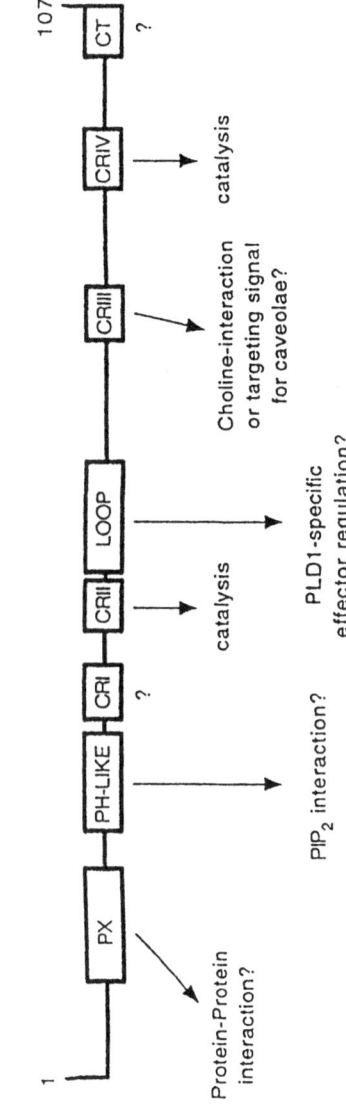

Fig. 4. Linear sequence of hPLD1 showing conserved regions (CR I–IV) and other motifs. Postulated functional roles for the various motifs are shown. From Sung et al. (1996b) by permission of the authors and publisher

3
Catalytic Mechanism

As described above, the HKD motifs are essential for catalysis, and structural and other studies show that they dimerize to form the catalytic center (Sung et al. 1997; Xie et al. 1998, 2000a). Experiments with $H_2{}^{18}O$ showed that PLD catalyzes bond cleavage of P–O rather than C–O (Holbrook et al. 1991), and an early scheme for the catalytic reaction proposed a ping-pong mechanism (Stanacev & Stuhne-Sekalec, 1970) with the formation of a covalent phosphatidyl-enzyme intermediate, with the free SH group of Cys as the phosphatidate acceptor (Yang et al.1967). However there is now strong evidence that His is the acceptor (Waite, 1999). In addition to being part of the HKD motif, His has been shown to be labeled with ^{32}Pi during experiments with Nuc or *Yersinia* murine toxin (Ymt) employing the phosphate-water exchange reaction that members of the PLD superfamily catalyze (Gottlin et al. 1998; Rudolph et al. 1999). This labeling was abolished by hydroxylamine and by mutation of either the conserved His or Lys, but was unaffected by mutation of the conserved Ser. Tungstate, which is a phosphate analogue and an inhibitor of the reaction, also binds to His, Lys and Asn in the active site of Nuc (Stuckey & Dixon, 1999), and phosphate is in close contact with His, Lys and Asn in *Streptomyces* PLD (Leiros et al. 2000).

A two-step catalytic mechanism has been proposed for PLD, based on studies of Nuc and Ymt (Fig. 5, Stuckey & Dixon, 1999; Rudolph et al. 1999). The first is a nucleophilic attack on the substrate phosphorus by the imidazole N of one of the active site His residues, with the second active site His acting as a general acid to donate a H^+ to the O of the leaving group. The covalent linkage between His and phosphate generates a phosphatidyl-enzyme intermediate.

Although the role of the HKD motifs in the catalytic mechanism is unquestioned, other regions of PLD are required for catalysis. One region is conserved sequence III and another is the extreme C-terminus, since limited

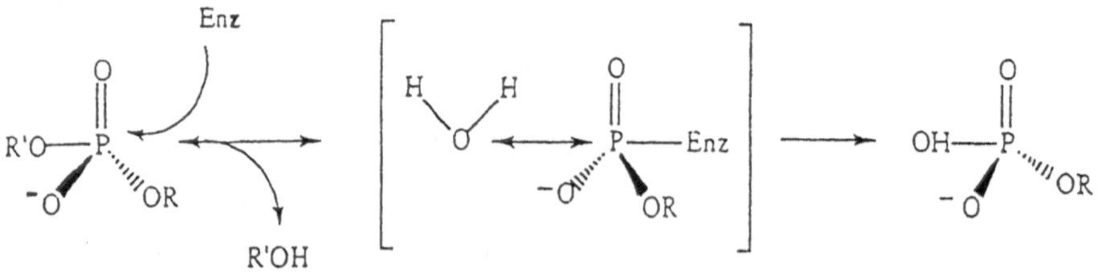

Fig. 5. Two step mechanism for catalysis by PLD involving a covalent enzyme intermediate. From Gottlin et al. 1998 by permission of the authors and publisher

Rat PLD1 **KEAIVPMEVWT**

Human PLD1 **KEAIVPMEVWT**

Rat PLD2 **KRGMIPLEVWT**

Human PLD2 **KEGMIPLEVWT**

C. elegans PLD **KEGLVPSAVFT**

Fig. 6. Sequences of C-termini   Drosophila PLD **KEGLIPTSVWT**
of PLDs from various species   Yeast SPO14 **SDRLSPMEIYN**

mutations of these sequences abolish catalysis (Sung et al. 1999b; Xie et al. 2000b; Liu et al. 2001). The C-terminal four amino acids are conserved in PLD1 and PLD2 and are similar to those in putative PLDs from *C. elegans* and *D. melanogaster*, but deviate from those in fungi, bacteria and plants (Fig. 6). The mammalian sequence is not required for association of the HKD motifs or for post-translational modifications such as palmitoylation or phosphorylation (Xie et al. 2000b, 2001). The activity of some forms of PLD1 with mutations in the C-terminus can be partially restored by co-expression of the C-terminal half of wild-type PLD (Xie et al. 2000b) or by addition of peptides corresponding to the C-terminus (Liu et al. 2001).

Since the phospholipid substrate of PLD is located in cell membranes, membrane association is critical for catalytic activity. Mutagenesis studies and sequence alignments have identified several sequences potentially involved in membrane interactions. One is the PH domain, which, in many proteins, is involved in binding to phosphoinositides in membranes. However, deletion of this domain in PLD does not abolish catalytic activity or membrane association (Park et al, 1998; Xie et al. 1998, 2000a; Sung et al. 1999b), implying the existence of other domains that target the enzyme to membranes. Another PIP_2 binding site has been identified in PLD1 (Sciorra et al. 1999). Mutation of this site abolishes activity, but does not alter membrane association.

4
Cellular Location of PLD1 and PLD2

Studies of the subcellular location of mammalian PLD isozymes have involved measurements of enzyme activity in subcellular fractions, Western blotting of subcellular fractions, immunocytochemistry and overexpression of PLD isozymes tagged with green fluorescent protein (GFP) or hemagglutinin (HA) in several cell types. Subcellular fractionation of liver demonstrated that the specific activity of PLD was highest in plasma membranes, Golgi and nuclei (Provost et al. 1996). Studies in other tissues or cells also revealed the presence of PLD activity in plasma membranes, Golgi and nuclei (Liscovitch et al. 1999). However, these reports did not define the distribution of PLD1 or PLD2. Because the affinity and specificity of most antibodies to these isozymes are generally not sufficient for use in subcellular fractionation or immunofluorescence studies, researchers have mainly relied on the expression of GFP- or HA-tagged PLD1 and PLD2 in cell types such as COS, RBL, CHO, 3Y1 and REF-52 cells. However, there has been one report of the intracellular localization of endogenous PLD1 in GH$_3$ pituitary and NRK kidney cells. This utilized highly sensitive antibodies together with immunofluorescence, immunogold electron microscopy and cell fractionation to localize PLD1 to the Golgi apparatus (Freyberg et al. 2001). Although the enzyme was concentrated in this organelle, it exhibited a diffuse reticular staining pattern with some localization in late endosomes and lysosomes. Disruption of the Golgi with brefeldinA (BFA) or 1-butanol caused release of PLD1. Interestingly, overexpression of PLD1 led to its mislocalization in a heterogenous population of small vesicles.

Other studies using GFP- or HA-tagged PLD1 have localized PLD1 to vesicular structures in the perinuclear region (Fig. 7), but have given variable results with respect to the plasma membrane or Golgi (Colley et al. 1997; Brown et al. 1998; Kim et al. 1999b; Sugars et al. 1999; Emoto et al. 2000; Lee et al. 2001; Kam and Exton 2001). Surprisingly, this approach has localized PLD2 primarily to the plasma membrane (Colley et al. 1997; Emoto et al. 2000; Lee et al. 2001) even though this isozyme appears less likely than PLD1 to be involved in signal transduction. Some studies have reported constitutive association of PLD2 and PLD1 with growth factor receptors (Slaaby et al. 1998; Min et al. 1998a). Western blotting has shown marked PLD1 immunoreactivity in the plasma membranes of chromaffin cells (Vitale et al. 2001) and 3Y1 fibroblasts (Kim et al. 1999b), with significant amounts also in the Golgi and endoplasmic reticulum fractions of the fibroblasts. Immunofluorescence and confocal microscopy studies have also shown that expressed PLD1 is localized to the periphery of chromaffin cells (Vitale et al. 2001). On the other hand, strong PLD2 immunoreactivity has been reported in cardiac sarcolemma (Park et al. 2000) and in the periphery of PC12 cells transfected with PLD1 (Lee et al. 2001). Recently PLD activity has been identified in caveolae, and both PLD1 and PLD2 have been reported to be present (Czarny et al. 1999, 2000; Iyer & Kusner, 1999; Sciorra & Morris 1999; Y. Kim et al. 1999, 2000). Some studies, but not all (Vitale et al. 2001), have reported translocation of PLD1 in response to cell stimulation (Brown et al. 1998; Emoto et al. 2000). However, because of the strong association of PLD with membranes, these changes probably involve the relocalization and/or

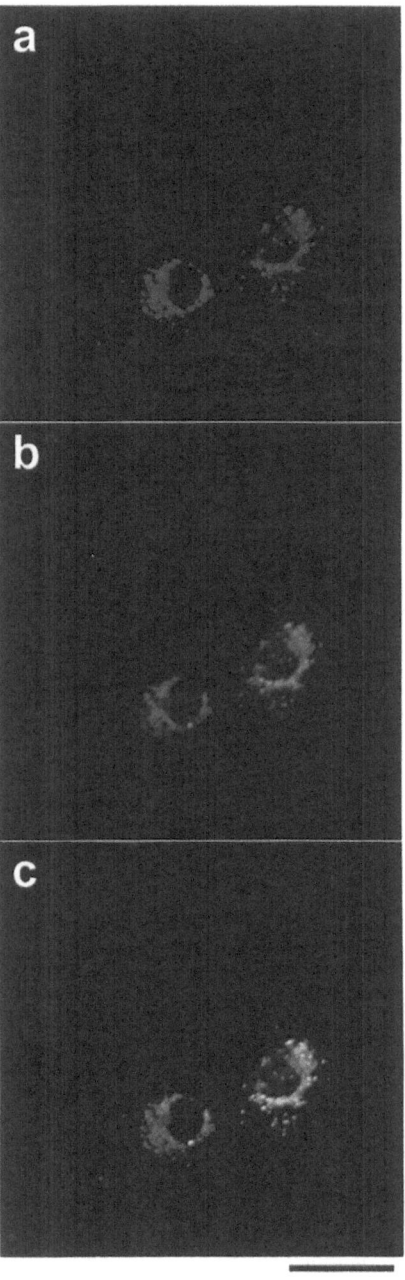

Fig. 7a-c. Cellular location of PLD1 in COS-7 cells. The cells were transfected with constructs for Xpress-tagged rPLD1 and EGFP-tagged rPLD1. Panel a shows the location of Xpress-tagged rPLD1 using a monoclonal antibody to Xpress. The EGFP fluorescence is shown in panel b, and the merged image is shown in panel c. From Kam & Exton (2001) by permission of the authors and publisher

40 µm

fusion of membranes rather than movements of PLD1 per se. A recent report has shown co-localization of PLD1 and PLD2 with the actin cytoskeleton in the periphery of COS7 and PC12 cells expressing these PLD isozymes
(Lee et al. 2001) and a functional interaction with this cytoskeleton has been
proposed. As noted in the next section, phosphorylation and palmitoylation
of PLD alter its association with membranes.

5
Posttranslational Modification of PLD

Several studies have demonstrated that PLD1 and yeast PLD (Spo14) are
phosphorylated on Ser/Thr residues in unstimulated cells (Rudge et al. 1998;
Y. Kim et al. 1999, 2000; Xie et al. 2000a, 2001). This results in the appearance of slower migrating bands on SDS polyacrylamide gel electrophoresis
(Fig. 8, Rudge et al. 1998; Xie et al. 2000a; Kim et al. 2000). These bands are
lost by treatment of the enzyme with alkaline phosphatase or with protein
phosphatases1and 2A, which are specific for phosphorylated Ser or Thr
residues (Rudge et al. 1998; Xie et al. 2000a). However, the endogenous
phosphorylation appears to have little effect on the catalytic activity of the
enzymes (Xie et al. 2000a). In contrast, it influences their subcellular localization (Fig. 8, Rudge et al. 1998; Xie et al. 2000a). The phosphorylated form
of PLD1 is found exclusively in membranes in COS7 cells in which it is over-

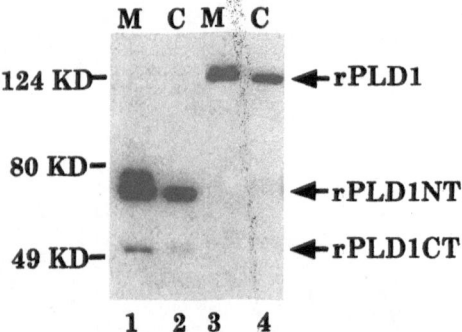

Fig. 8. Phosphorylated PLD1 is associated with membranes. COS-7cells co-
expressing the N- and C-terminal halves of rPLD1 or full-length rPLD1 were frac-
tionated into membrane (M) and cytosol (C) fractions and Western blotted with
antibodies against these proteins. The upper bands (slower migration on SDS polyacrylamide gel electrophosis) seen in the membrane fractions represent the phosphorylated enzyme. From Xie et al. (2000a) by permission of the authors and publisher

expressed (Xie et al. 2000a; Kim et al. 2000). Phosphorylation of Spo14 is required for its relocalization during meiosis, where it participates in spore formation (Rudge et al. 1998). The nature of the protein kinase(s) responsible for the endogenous phosphorylation of PLD1 and Spo14 is unknown. There has been one report that PLD2 is phosphorylated on Ser/Thr residues and that this inhibits its activity (Watanabe & Kanaho 2000). Modification of PLD isozymes by tyrosine kinases and PKC will be discussed below in the relevant sections.

PLD1 has been reported to be palmitoylated when expressed in COS-7 cells (Manifava et al. 1999; Sugars et al. 1999; Xie et al. 2001). This occurs on Cys residues 240 and 241 in hPLD1 and rPLD1 (Sugars et al. 1999; Xie et al. 2001). The effect of palmitoylation on the catalytic activity of PLD1 is un-

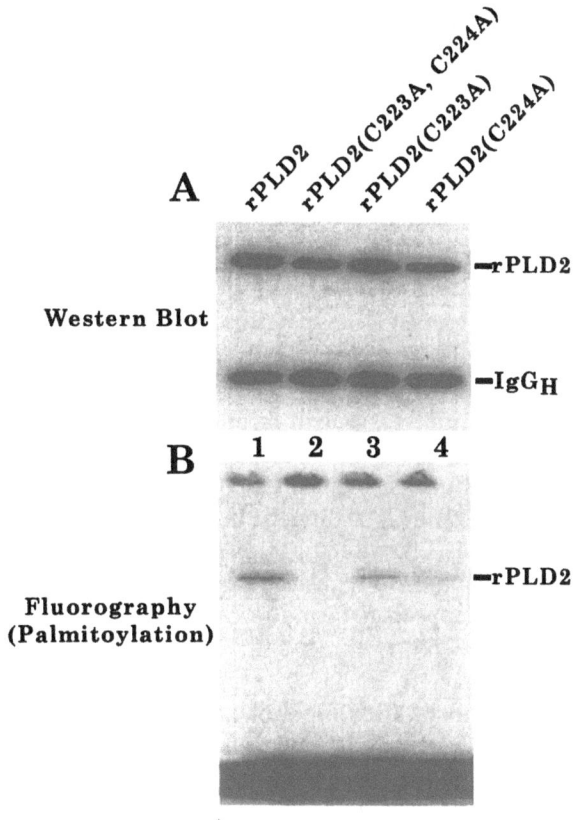

Fig. 9A, B. PLD2 is palmitoylated on cysteines 223 and 224. The incorporation of [³H]palmitate into rPLD1 and the alanine mutants is shown by fluorography. Palmitoylation was totally lost in the C223A, C224A double mutant. From unpublished studies by Z. Xie, W.-T. Ho and J.H. Exton

clear. Although palmitoylation is eliminated when the two Cys residues are mutated to Ala, the enzyme is fully active in vitro (Sugars et al. 1999). When expressed in COS-7 cells, there is a partial loss of catalytic activity (Sugars et al. 1999; Xie et al. 2001), but it is not clear that this is due to the loss of palmitoylation or to the Cys mutations per se. This is because mutant forms of PLD1 in which the palmitoylation sites have been deleted show high catalytic activity (Xie et al. 2000). Loss of palmitoylation does, however, weaken the association of PLD1 with membranes (Xie et al. 2000) and causes its relocalization within COS-7 cells (Sugars et al. 1999). rPLD2 is also palmitoylated when expressed in COS-7 cells (Fig. 9) and the modification is eliminated by mutations of Cys 223 and Cys 224 (corresponding to Cys 240 and 241 in rPLD1). Although this results in a loss of basal catalytic activity, the activity in the presence of phorbol ester is not decreased as seen also for PLD1. As in the case of PLD1, mutations of the palmitoylation site reduces the membrane association of PLD2.

There has been one report that PLD1 expressed in Sf9 cells is glycosylated and that this determines its distribution between the membrane and soluble fractions (Min et al. 1998b). There have been no reports of this post-translational modification in mammalian cells. Neither PLD1 nor PLD2 has the consensus sequences for modification by myristoylation or prenylation.

6
Regulation of Phospholipase D

The PLD activity of many cell types is increased by a variety of hormones, neurotransmitters, growth factors and cytokines (Exton 1997; Exton 1999). Many of the agonists act through membrane receptors coupled to hetero-trimeric G proteins e.g. G_q, G_i, G_{13}, but there is no evidence that these G proteins activate PLD directly. In contrast, they act on the enzyme via signaling cascades. Likewise, the tyrosine kinase activity of growth factor receptors does not appear to regulate PLD by *direct* phosphorylation and the regulation again appears to be indirect.

6.1
Role of Phosphatidylinositol 4,5 Bisphosphate

Many factors have been shown to affect PLD activity directly. One factor that is essential for catalytic activity of many isozymes in animals, plants and protozoa is PIP_2 (Brown et al. 1993; Liscovitch et al. 1994; Hammond et al. 1995; Pappan et al. 1997; Wang et al. 2001). PIP_3 is also effective and phos-

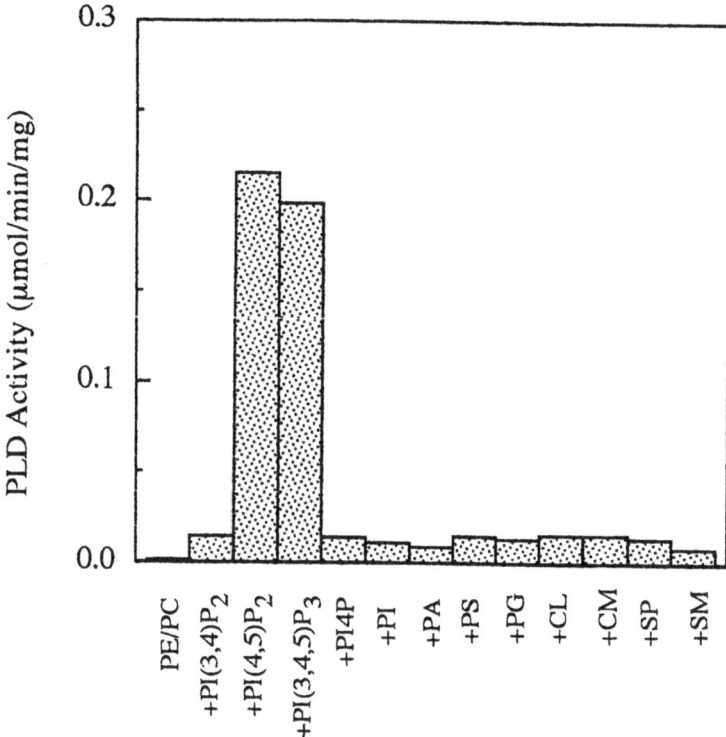

Fig. 10. PLD1 activity is stimulated in vitro by PI 4,5-P_2 and PI 3,4,5-P_2, but not by other phospholipids. In vitro PLD assays were carried out with recombinant rPLD1 in the presence of myristoylated ARF3 and GTPγS. Where indicated PIP_2 was replaced by other lipids in the substrate PL vesicles by other lipids. From Min et al. (1998b) by permission of the authors and publisher

phatidylinositol 4-phosphate (PIP) can have an effect, but all other phospholipids are ineffective (Fig. 10, Brown et al. 1993; Liscovitch et al. 1994; Schmidt et al. 1996b; Pappan et al. 1997; Hammond et al. 1997; Min et al. 1998b; c.f. Hodgkin et al. 2000). A binding site for PIP_2 on PLD2 has been located between conserved sequences II and III by the use of a photoreactive PIP_2 derivative, and by the demonstration that the site contains amino acid residues that are required for PLD activity (Sciorra et al. 1999). The sequence contains multiple residues that are conserved between PLD1, PLD2 and Spo14. Although mutation of these residues in PLD2 and Spo14 causes loss of PIP_2 binding and catalytic activity, it does not affect the cellular localization of these enzymes, indicating that other sequences are involved in membrane association (Sciorra et al. 1999). Another study has localized another

PIP$_2$ binding site in PLD1 to the isolated PH domain using surface plasmon resonance (Hodgkin et al. 2000). Mutagenesis of conserved residues in this domain caused a loss of PLD activity. However, these results must be interpreted with caution since another study showed that deletion of the PH domain produced little effect on activity (Hoer et al. 2000) and other studies have shown that deletion of the N-terminal 319 residues of PLD1 or PLD2, which contain the PH domain, results in an enzyme that is still active (Park et al. 1998; Xie et al. 1998; Sung et al. 1999a,b).

Evidence that PIP$_2$ is required for the activity of PLD in vivo has come from several studies utilizing permeabilized cells. These have included examinations of the effects of neomycin, which binds PIP$_2$ and inhibits PLD activity stimulated by MgATP and GTPγS or by PKC and PMA (Liscovitch et al. 1994; Pertile et al. 1995; Ohguchi et al. 1996; Schmidt et al. 1996b). The *Clostridium difficile* Toxin B which inactivates Rho proteins reduces PIP$_2$ levels in cells (Schmidt et al. 1996b). This is because Rho stimulates PIP$_2$ synthesis through its action on PI 4-P 5-kinase via Rho-kinase (Oude Weernink et al. 2000). The decrease in PIP$_2$ is associated with decreased membrane-associated PLD activity and this can be reversed by adding PIP$_2$ to the membranes (Schmidt et al. 1996b). Both Toxin B and the C3 exoenzyme of *C. botulinum*, which also inactivates Rho, inhibited the ability of carbachol or AlF$_4^-$ to activate PLD in HEK cells expressing the M3 muscarinic cholinergic receptor (Schmidt et al. 1996a). However, these results can also be explained by inhibition of the direct effect of Rho on PLD (see below).

Another line of support for the dependence of PLD on PIP$_2$ for activity comes from studies of the interaction and co-localization of Type Iα PI 4-P 5-kinase and PLD (Divecha et al. 2000). These have utilized co-expression of PLD2 and the kinase in COS-7 cells and have shown by co-immunoprecipitation that the two proteins associate. Furthermore, they co-localized in a submembranous vesicular compartment whenexpressed in PAE cells. Local control of the activity of PLD2 by PIP$_2$ was indicated by the finding that PLD activity was reduced when a kinase-dead mutant of PI 4-P 5-kinase was transfected in place of the wild-type enzyme (Divecha et al. 2000). Control of PLD activity by changes in the level of PIP$_3$ seems unlikely in view of the much higher level of PIP$_2$ in cells. Accordingly, concentrations of wortmannin that greatly decrease the level of PIP$_3$ are without effect on basal and GTPγS-stimulated PLD activity in permeabilized HL-60 cells (El Hadj et al. 1999).

6.2
Role of Protein Kinase C

PKC is an important mediator of agonist activation of PLD in some cell lines, but may play a minimal role in other cell types. In most cells, treatment with 4β-phorbol 12-myristate 13- acetate (PMA) or related phorbol esters stimulates PLD markedly, but agonist effects are often of less magnitude (Fig. 11A). The role of PKC in the effects of agonists has generally been explored by testing the effects of inhibitors of the kinase action of PKC. These have included a variety of non-specific and specific inhibitors e.g. staurosporine, H-7, sphingosine, Ro-31-8220, bisindolylmaleimide I, Gö6076, chelerythrine, bryostatin 1 and calphostin C. Most of these inhibitors target the ATP-binding site of PKC, whereas sphingosine, bryostatin1 and calphostin C interact at the DAG/phorbol ester binding site (Nixon 1997). They generally inhibit the effects of PMA on PLD almost completely, but have partial or, in some cases, no effects on natural agonists (Exton 1997), implying the existence of other mechanisms of PLD activation not involving PKC.

Other approaches have been used to define the role of PKC in agonist activation of PLD. A common one is prolonged treatment of cells with phorbol ester to down-regulate PKC. This procedure results in a time-dependent loss of the Ca^{2+}-dependent and -independent typical PKC isozymes, but no loss of the atypical isozymes (Kiley et al. 1990; Olivier and Parker 1992). Except in a few cases, PKC down-regulation results in complete or partial inhibition of the effects of growth factors or G protein-linked agonists on PLD (Exton 1997). The usual mechanism by which these factors and agonists activate PKC is by stimulating the hydrolysis of PIP_2 to generate DAG. This involves the β-isozymes of phospholipase (PLC) in the case of agonists that activate G_q and G_i, and the γ-isozymes of PLC in the case of growth factors that induce autophosphorylation of their receptors. In the case of platelet-derived growth factor (PDGF), the evidence for the involvement of PLC is very strong. For example, in cells expressing mutant receptors that are unable to induce activation of PLC, PDGF was incapable of activating PLD (Fig. 12, Yeo et al. 1994). On the other hand, the PLD response was restored with receptors that mediate PLC activation, but do not elicit other PDGF signaling pathways (Yeo et al. 1994). In another study utilizing fibroblasts overexpressing PLC-γ, the PLD response to PDGF was greatly enhanced, except when tyrosine kinase activity was inhibited (Lee et al. 1994). Overexpression of the α- and β1-isozymes of PKC also enhanced the PLD responses to PMA, G protein-linked agonists or PDGF (Pai et al. 1991; Pachter et al. 1992; Eldar

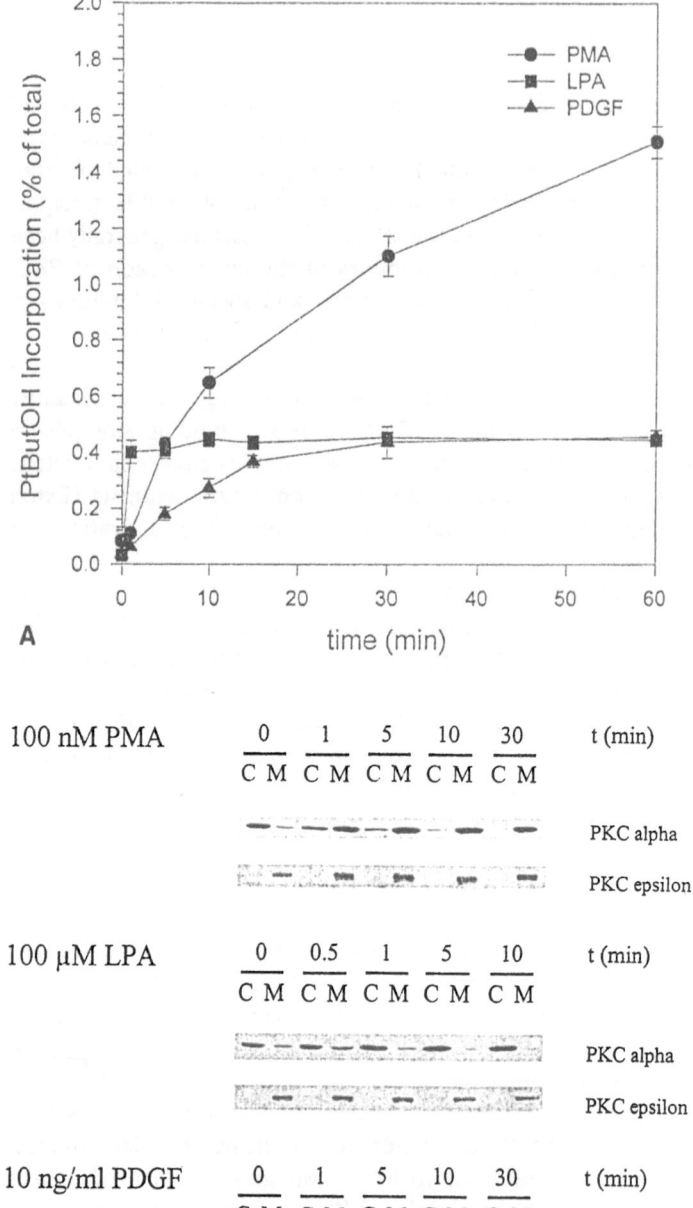

A

100 nM PMA   0   1   5   10   30   t (min)

C M C M C M C M C M

PKC alpha

PKC epsilon

100 µM LPA   0   0.5   1   5   10   t (min)

C M C M C M C M C M

PKC alpha

PKC epsilon

10 ng/ml PDGF   0   1   5   10   30   t (min)

C M C M C M C M C M

PKC alpha

B   PKC epsilon

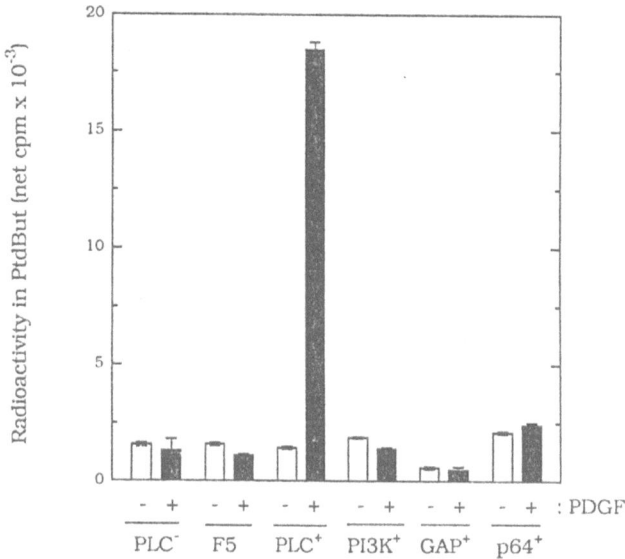

Fig. 12. Activation of PLD by PDGF in TRMP kidney epithelial cells expressing mutated PDGF receptors. Cells were labeled with [³H]myristic acid and PLD measured by [³⁺H]PtdBut formation. Mutant receptors were: F5, lacking all tyrosines required for signaling; PLC-lacking the tyrosine required for coupling to PLC; PLC⁺, having only the tyrosine for coupling to PLC. The other mutants (PI3K⁺, GAP⁺, p64⁺) having the tyrosines for coupling to other signals. From Yeo et al. by permission of the authors and publisher

et al. 1993), while depletion of PKC-α by antisense methods decreased activation of the enzyme (Balboa et al. 1994).

An interesting approach to exploring the role of PKC isozymes in regulating PLD has involved the use of a peptide which binds to the receptor for activated PKC (RACK1) (Thorsen et al. 2000). Cell lines expressing this peptide showed inhibition of cell functions mediated by conventional (Ca²⁺-dependent) isozymes of PKC. Thus these cell lines showed a great loss of PMA-stimulated PLD activity, whereas the parental cells did not. On the other hand, the peptide-expressing cells showed normal uptake of choline

Fig. 11A, B. Correlation of activation of PLD with membrane translocation of PKCα. A. Time courses of activation of PLD in NIH3T3 fibroblasts by PMA, LPA and PDGF. Cells were labeled with [³H]myristic acid and PLD measured by [³H]PtdBut formation. B. The presence of PKCα and PKCε in the cytosolic (C) and membrane (M) fractions at various times was determined by Western blotting. From unpublished studies by F.G. Buchanan, M. McReynolds and J. H. Exton

and its incorporation into PC (Thorsen et al. 2000). These data provide strong evidence for a role of α-and β-isozymes of PKC in PMA activation of PLD, possibly involving RACK1.

Activation of PKC isozymes by agonists is usually associated with their translocation from the cytosol to the plasma membrane and other intracellular membranes (Jaken 1997) The movement to the plasma membrane occurs presumably because the DAG generated by PIP_2 breakdown remains in the plasma membrane until it is metabolized. Thus the simplest scheme for the activation of PLD by agonists that stimulate PLC is that there is translocation of typical PKC isozymes to the plasma membrane where they interact with PLD to cause its activation and possibly phosphorylation. In support of this scheme, studies of the relationship between PMA- or agonist-induced membrane translocation of PKC isozymes and PLD activation have shown a good correlation (Fig. 11A & B, Kim et al. 1999b). One of these studies identified plasma membranes as the site of PKCα-PLD1 interaction (Kim et al. 1999b) and more recently, caveolae have been proposed to be the site (Kim et al. 1999b, 2000).

As described in detail below, deletion of the first 319 or 325 amino acids of PLD1 yields an enzyme that is unresponsive to PKC, but is still activated by Rho or ARF (Park et al. 1998; Sung et al. 1999b). Insertion of 5 amino acids at Glu 87 of PLD1 also renders the enzyme unable to be activated by PKC, but it remains normally responsive to Rho and ARF (Zhang et al. 1999). Expression of N-terminally truncated PLD1 in COS-7 cells results in a PLD activity that is no longer responsive to constitutively active $G\alpha_q$ which is a G protein α-subunit that activates PLC and hence PKC (Exton 1996). Likewise, cells expressing PLD1 with an insert at Glu87 show greatly impaired responses to agonists that activate the M1-and M3-muscarinic and bombesin receptors (Zhang et al. 1999) which are known to couple to G_q (Exton 1996). These results with mutant PLD1 enzymes strongly support a role for PKC in the activation of PLD1 by agonists whose receptors are linked to G_q.

A major issue in studies of the activation of PLD by PKC is the role of phosphorylation. As noted below, there is strong evidence that the α- and β-isozymes of PKC can directly activate PLD1 in vitro by a non-phosphorylation mechanism (Fig.13, Conricode et al. 1992, 1994; Singer et al. 1996; Hammond et al. 1997; Min et al. 1998; Sciorra et al. 2001). However, numerous studies with inhibitors of the kinase activity of PKC have shown inhibition of the activation of PLD by PMA and agonists in many cell lines (Exton, 1997). Some investigators have also reported phosphorylation of PLD1 in response to PMA in COS or Sf9 cells overexpressing this enzyme (Min &

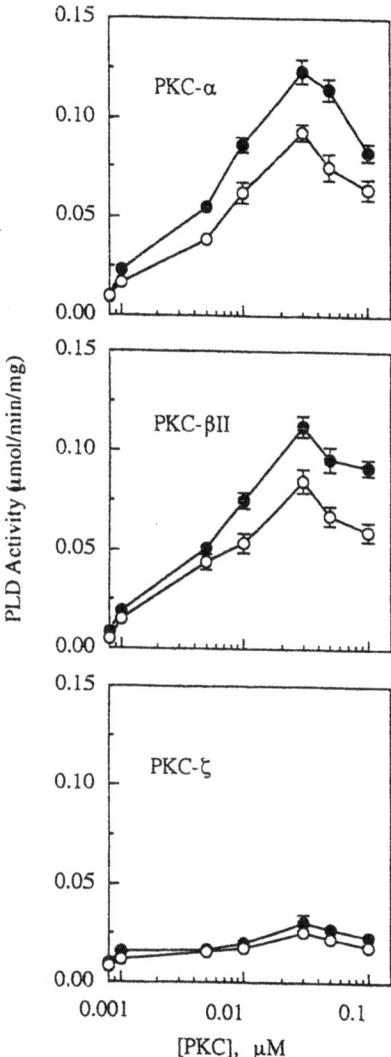

Fig. 13. PKCα and PKCβ directly activate PLD1. Increasing concentrations of recombinant PKCα and PKCβII were incubated with recombinant rPLD1 and PLD activity measured in vitro in the absence of ATP, with (•) and without (o) PMA. PKCζ and all other PKC isozymes (not shown) were ineffective. Adapted from Min et al. (1998b) by permission of the authors and publisher

Exton 1998; Y. Kim et al. 1999, 2000). However, there has been only one report of PMA-stimulated Ser/Thr phosphorylation of *endogenous* PLD in cells (Y. Kim et al. 1999). Furthermore, although PLD1 can be phosphorylated by PKCα in the presence of ATP in vitro (Min et al. 1998b) its activity is actually inhibited (Hammond et al. 1997; Min et al. 1998b). Three residues (Ser 2, Thr 147 and Ser 561) have been identified as sites of phosphorylation on PLD1 by PKCα (Kim et al. 2000) and single or triple mutations of these residues to Ala reduced the activation by PMA in COS cells to approximately

50% of that seen with wild type PLD1 (Y. Kim et al. 1999, 2000). Since the mutations abolished the phosphorylation of PLD1 induced by PMA in COS cells overexpressing this enzyme, as indicated by the disappearance of slower migrating bands on polyacrylamide gel electrophoresis (Kim et al. 2000), these findings suggest that both phosphorylation and non-phosphorylation mechanisms of activation of PLD were operative in these cells.

Concerning the in vivo effects of PKC inhibitors which block the kinase activity by interacting at the ATP-binding site in the catalytic domain, there are several possible explanations. One obvious explanation is that these compounds block the phosphorylation of PLD by PKC. However, although it has been shown that mutation of the phosphorylation sites of PKC on PLD1 partially blocks the activation of the enzyme by PMA in intact cells (Kim et al. 2000), it has not been shown that these mutations inhibit the ability of PKCα to activate PLD1 in vitro. As noted above, PKC-induced phosphorylation of the PLD1 in vitro actually causes a decrease in activity (Hammond et al. 1997; Min et al. 1998b). Thus the relationship between the phosphorylation of PLD by PKC and its activation by the kinase remains unclear.

Another explanation is that another protein(s) that regulates PLC activity is phosphorylated by PKC. Pretreatment of plasma membranes from neutrophils with PKC, ATPγS and PMA to phosphorylate endogenous proteins resulted in stimulation of subsequently added PLD (Lopez et al. 1995). The effect was blocked by staurosporine, required Ca^{2+}, and was observed with the α-, β1- and γ- isozymes of PKC, but not the Ca^{2+}-independent and atypical isozymes. The nature of the phosphorylated PLD regulator was not defined.

A third possibility is that PKC inhibitors block the interaction between PKC and PLD. Against this hypothesis are studies of PLD activation by PKCα in vitro which reported no inhibition by staurosporine (Singer et al. 1996) and showed, by cleavage of PKCα by trypsin into its regulatory and catalytic domains, that the regulatory domain alone could activate PLD (Singer et al. 1996). However, it was not clear whether or not the holoenzyme was more effective than the regulatory domain, i.e. whether or not the catalytic domain also played some role in the interaction of PKCα with PLD. Although most studies indicate that PKC interacts with the N-terminal 325amino acids of PLD (Park et al. 1998; Xie et al. 1998; Min & Exton 1998; Sung et al. 1999b), there is some evidence from co-immunoprecipitation studies of additional interaction sites (Min & Exton 1998; Sung et al. 1999b). These could possibly involve the catalytic domain of PKC. In support of this idea, unpublished studies in the author's laboratory have shown that Ro-31-8220 and bisindolylmadeimide inhibit the interaction of endogenous PKCα with endogenous PLD in NIH3T3 cells.

In addition to the in vivo evidence for a role of PKC in PLD regulation, many in vitro studies have shown direct activation of PLD1 by PKC isozymes (Fig. 13, Singer et al. 1996; Hammond et al. 1997; Min et al. 1998b; Sciorra et al. 2001). Several interesting features have arisen from these studies. The first is that ATP is not required for the effect and that inhibitors of PKC kinase activity are without effect, thus confirming earlier reports of a non-phosphorylation mechanism for PLD activation by PKC (Conricode et al. 1992, 1994). The second is that the response is elicited only by the α- and β-isozymes of PKC (Conricode et al. 1994; Ohguchi et al. 1996; Min et al. 1998b; Sciorra et al 2001). The conclusion that the kinase activity of PKC is not needed for the in vitro activation of PLD is supported by the observation that the regulatory domain of PKC alone is capable of activating the enzyme, although it is less effective than the holoenzyme (Singer et al. 1996; Sciorra et al. 2001). Surprisingly, addition of ATP to incubations of recombinant PLD1 with PKC-α results in phosphorylation and inhibition of the phospholipase (Hammond et al. 1997; Min et al. 1998b). These data indicate that direct phosphorylation of PLD by PKC in vitro does not lead to activation. Whether or not the phosphorylation leads to alterations in the effects of activators or inhibitors of the enzyme remains to be determined. In contrast to PLD1, PLD2 is weakly or not activated by PKC in vitro (Colley et al. 1997; Slaaby et al. 2000; Sciorra et al. 2001), although it responds well to PMA when expressed in COS-7 or Sf9 cells (Siddiqi et al. 2000; Slaaby et al. 2000). This implies that PKC can activate the enzyme in vivo through indirect mechanisms. In HEK cells transfected with cDNAs for the insulin receptor and PLD2 or PLD1, insulin stimulation of PLD activity was enhanced by cotransfection with PKCα plus PLCγ (Slaaby et al. 2000). Co-immuno-precipitation studies also showed a physical association between PLD2 and PKCα.

Mutagenesis and binding studies have located a major site of interaction of PLD1 with PKCα in the N-terminal sequence of the phospholipase. Thus, deletion of the first 319 or 325 amino acids of PLD1 results in an enzyme that is no longer responsive to PKCα in vitro or PMA in vivo (Fig. 14, Park et al. 1998; Sung et al. 1999b; Xie et al. 1998). Binding studies using PLD sequences fused to GST have also shown strong interaction of PKC(with the first 318 amino acids of PLD1 (Min & Exton 1998). However, the protein kinase could also associate with a mutant PLD1 in which the first 325 amino acids were deleted (Sung et al. 1999b), indicating that there was an additional interaction site(s).

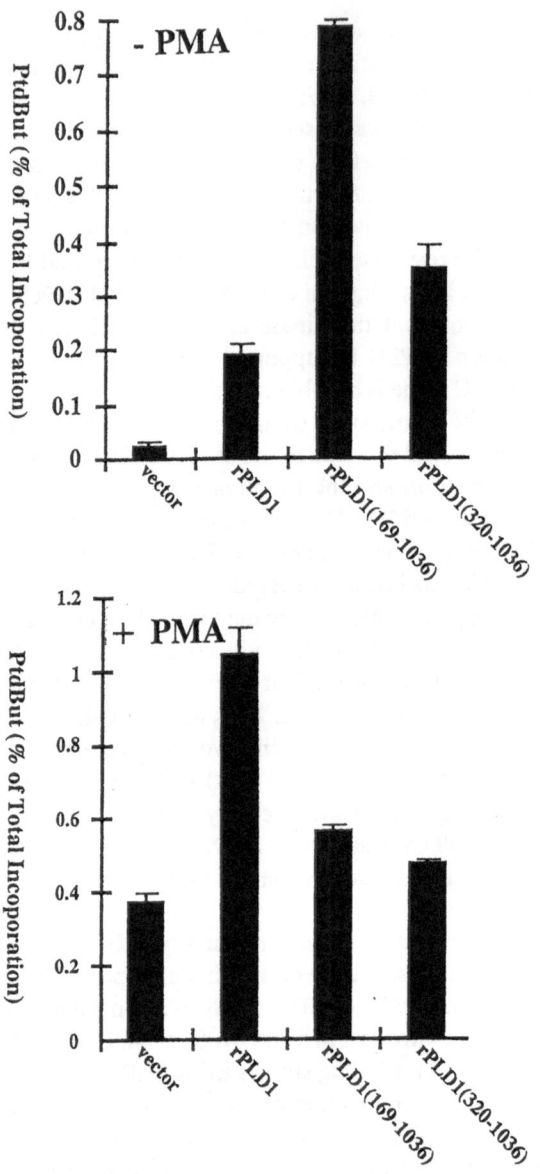

Fig. 14. N-terminal deletion mutants of rPLD1 have increased basal activity, but do not respond to PKC. Wild-type and N-terminally truncated forms (169–1036 and 320–1036) of rPLD1 were expressed in COS7 cells which were labeled with [$^{3+}$H]myristic acid for measurement of PLD activity by [^{3}H]PtdBut formation. The figure shows that both truncated enzymes had increased basal activity but totally lost their response to PMA (compare values in the presence and absence of PMA). Adapted from Xie et al. 1998 with permission of the authors and publisher

Another feature that has emerged from in vitro studies of PLD1 is that PKCε interacts synergistically with RhoA (or Cdc42) and ARF to activate the enzyme (Singer et al. 1996; Ohguchi et al. 1996; Hammond et al. 1997; Hodgkin et al. 2000). This synergism is quite striking, and is also observed when Rho (or Cdc42) and ARF are combined. The mechanistic basis for these synergisms is unknown. The synergism between PKC and RhoA could explain why the stimulation of PLD by phorbol esters in some cell types is partly inhibited by inactivation of RhoA (Malcolm et al. 1996; Senogles 2000).

Although the conventional isozymes of PKC are usually stimulatory to PLD, the situation with PKC(in vitro is unclear. An early report indicated that this isozyme mediated the effect of extracellular ATP on PLD in mesangial cells (Pfeilschifter & Huwiler 1993). A role for PKCε was later supported by a study of the effects of D2 dopaminergic stimulation of PLD in GH4 pituitary cells (Senogles 2000). On the other hand, stable overexpression of PKCε and its regulatory domain in fibroblasts was observed to inhibit the effects of PMA and PDGF on PLD activity (Kiss et al. 1999). However, caution must be used in interpreting this type of study since a high level of expression of PKCε may induce effects that are not seen with endogenous levels of the enzyme.

A recent report has shown that the PKC-related protein kinases PKNα and PKNβ can interact with PLD1 as shown by co-immunoprecipitation studies in COS7 cells (Oishi et al.2001). The binding site was localized to residues 228–598 of PLD1. PKNα stimulated the activity of the enzyme, whereas PKNα had a modest effect. Another Ser/Thr kinase (casein kinase 2-like serine kinase) has been reported to phosphorylate PLD1, with Ser^{911} being one site of phosphorylation (Ganley et al. 2001). However, no change in in vitro catalytic activity was observed.

6.3
Role of Rho Family GTPases

Studies of the in vitro regulation of purified or membrane-associated PLD first demonstrated that the enzyme could be directly stimulated by members of the Rho and ARF families of small GTPases (reviewed in Exton 1997, 1999). It was subsequently shown that the PLD1 isozyme responded to Rho GTPases, but the PLD2 isozyme did not (Hammond et al. 1997; Colley et al. 1997; Lopez et al. 1998; Min et al. 1998b; Kodaki & Yamashita 1997). All members of the Rho family that were tested (RhoA, RhoB, Rac1, Rac2,

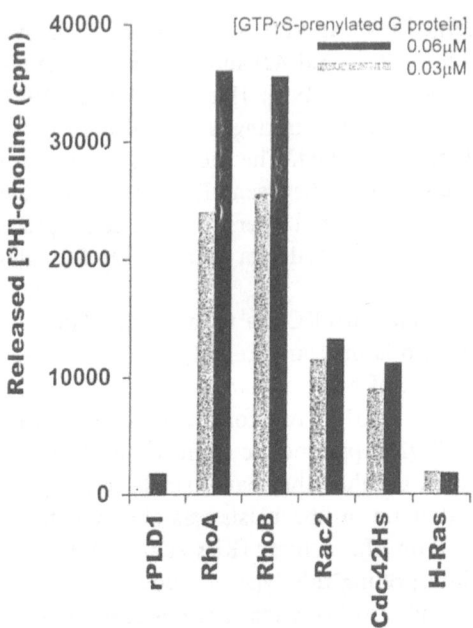

Fig. 15. Effects of different Rho family members on PLD1 activity. Recombinant rPLD1 was incubated with two different concentrations of GTPγS-liganded preny-lated Rho proteins and H-Ras. PLD activity was measured by the release of [^{3}H]choline from [^{3}H]choline-labeled PC. From Bae et al. (1998) by permission of the authors and publisher

Cdc42Hs) stimulated PLD1, but with differing efficacies (Fig. 15, Hammond et al. 1997; Bae et al. 1998). Several studies have localized the interaction site for RhoA to the C-terminal sequence of PLD1 (Sung et al. 1997; Yamazaki et al. 1999; Du et al. 2000b; Cai & Exton 2001). Thus active V^{14}RhoA was found to bind to a C-terminal fragment of hPLD1 including amino acids 663–1074 or 674–1074 using the yeast 2-hybrid system (Sung et al. 1997; Yamazaki et al. 1999). A similar, but shorter sequence in rPLD1 was found to bind acti-vated RhoA utilizing phage display (Cai & Exton 2001). These findings were reinforced by co-immunoprecipitation or "pull down" studies, which showed that the interaction was GTP-dependent (Yamazaki et al. 1999; Cai & Exton 2001). Peptides corresponding to the C-terminal sequence of PLD1 also blocked the ability of RhoA to stimulate the catalytic activity of the enzyme (Yamazaki et al. 1999; Cai & Exton 2001). Mutagenesis studies have revealed that certain residues are required for the interaction. One study showed that mutations in four amino acids in the Lys^{946} to Lys^{962} sequence in rPLD1 inhibited the activation of rPLD1 by V^{14}RhoA, but not PMA, in COS7 cells, and blocked the binding of the enzyme to GTPγS-liganded RhoA as shown by co-immunoprecipitation studies (Cai & Exton 2001). Another study indicated the importance of Ile[870] in hPLD1 (corresponding to Ile[882] in rPLD1) for binding to Rho in the yeast split-hybrid system and for activation of PLD (Du et al. 2000b). This group identified two other residues, but these required double mutations to see effects. These findings suggest that multiple residues in the C-terminus of PLD1 are involved in the RhoA interaction.

One study has examined the amino acid residues in Rho proteins required for interaction with PLD1 (Bae et al. 1998). This showed that RhoA and RhoB were more effective than Rac2 and Cdc42Hs in activating PLD1 in vitro in accord with Hammond et al. (1997). Experiments utilizing RhoA/Ras and RhoA/Cdc42 chimeras and mutagenesis revealed that residues Tyr[34], Thr[37] and Phe[39] in the activation loop of RhoA were required for stimulation of the phospholipase, but that other residues (Gln[52] and Asp[76]) determined the greater efficacy of RhoA compared with Cdc42Hs (Bae et al. 1998). Another study identified the insert loop of Cdc42 (amino acids 120–139) as being required for PLD1 activation (Walker et al. 2000). Since both RhoA and Cdc42 contain this insert, the results of the two groups are not contraditory. As described above, RhoA and Cdc42 can interacts synergistically with ARF and PKCα to activate PLD1 in vitro.

Rho family members have also been shown to be involved in the regulation of PLD in vivo. Thus expression of wild-type or constitutively active V[14]RhoA or V[12]Rac1 in fibroblasts or COS cells increases PLD activity (Hess et al. 1997; Park et al. 1997; Zhang et al. 1999; Du et al. 2000b; Cai & Exton 2001) and dominant negative N[19]RhoA or N[17]Rac1 attenuates the activation of PLD induced by EGF, PMA or constitutively active Gα₁₃ (Hess et al. 1997; Plonk et al. 1998; Meacci et al. 1999). Another commonly used approach to reduce RhoA activity in cells is by treatment with the C3 exoenzyme of *Clostridium botulinum* which inactivates RhoA by ADP-ribosylation (Aktories et al. 1989). Toxin B from *C. difficile* which inactivates Rho, Rac and Cdc42 through monoglucosylation (Just et al. 1995) has also been employed. The C3 exoenzyme has been shown to attenuate PLD responses to agonists in several cell lines (Fig. 16, Malcolm et al. 1996; Hess et al. 1997; Plonk et al. 1998; Meacci et al. 1999; Senogles 2000; Murthy et al. 2001) and to inhibit GTPγS stimulation of PLD in membranes (Kuribara et al. 1995; Schmidt et al. 1996a). Toxin B from *C. difficile* has similar effects to the C3 exoenzyme on carbachol-stimulated PLD activity in HEK cells expressing M3 muscarinic receptors (Schmidt et al 1996a,b). It also inhibits IgE receptor-coupled PLD activation in RBL-2H3 cells and suppresses activation of the enzyme by GTPγS and PMA in the permeabilized cells (Ojio et al. 1996).

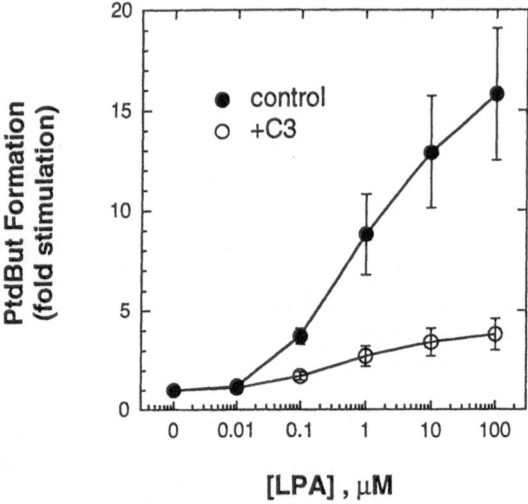

Fig. 16. Effect of C3 exoenzyme from *C. Botulinum* on the stimulation of PLD by LPA in Rat1 fibroblasts. PLD activity was measured by [^{3}H]PtdBut formation in cells labeled with [$^{3+}$H]myristic acid. Cells were scrape-loaded with C3 exoenzyme. From Malcolm et al. (1996) by permission of the authors and publisher

The in vivo effects of N^{19}RhoA and the clostridial toxins can be attributed to direct inhibition of the stimulatory effects of Rho protein on PLD1. This is supported by the observations that mutations in PLD1 that block binding of RhoA also abolish the activation of the enzyme by V^{14}RhoA in vivo. (Du et al. 2000; Cai & Exton 2001). However, indirect mechanisms for the regulation of PLD by Rho have been proposed. These include changes in PIP$_2$ (Schmidt et al. 1996b) since *C. difficile* Toxin B lowers both PIP$_2$ levels and PLD activity in HEK cells and membranes, and direct addition of PIP$_2$ to the membranes restores activity. Furthermore, several studies have shown that Rho can control the activity of PI-4P 5-kinase (Chong et al. 1994; Ren et al. 1996; Oude Weernink et al. 2000). Although these studies show that a decrease in PIP2 can reduce PLD activity in vivo, they have not shown that agonists stimulate the enzyme by increasing the level of this lipid.

Another postulated indirect mechanism involves Rho kinase (Schmidt et al. 1999). The evidence is that transfection of wild-type or constitutively active Rho kinase into HEK cells expressing M3 muscarinic receptors enhanced the stimulatory effect of carbachol on PLD activity without changing the effect of PMA. In contrast, kinase-deficient Rho kinase was without effect (Schmidt et al. 1999). However, active Rho kinase alone had minimal effects in the intact cells. In contrast, active Rho kinase increased basal and

MgATP plus GTPγS-stimulated PLD activity in HEK cell membranes and a Rho kinase inhibitor reduced the stimulatory effect of RhoA plus GTPγS and MgATP (Schmidt et al. 1999). The Rho kinase inhibitor also decreased the stimulatory effect of carbachol in the cells. Although these experiments point to a significant role for Rho kinase in M3 muscarinic activation of PLD, they do not clearly distinguish between Rho kinase effects on PIP2 synthesis vs. PLD activity.

Further analysis of the effects of activation of the M3 muscarinic receptor on PLD activity using overexpression of certain G protein α-subunits indicated that $G\alpha_{12}$ and $G\alpha_{13}$, but not $G\alpha_q$ enhanced the effect of carbachol, but not PMA, on the enzyme (Rümenapp et al. 2001). As expected, $G\alpha_{12}$ and $G\alpha_{13}$ did not influence the activation of PLC, whereas $G\alpha_q$ enhanced the effect of carbachol on this enzyme. Overexpression of two regulators of G protein signaling (RGSs) that were specific suppressors of either $G\alpha_q$ or $G\alpha_{12}/G\alpha_{13}$ signaling also indicated that cholinergic stimulation of PLD involved $G\alpha_{12}/G\alpha_{13}$ but not $G\alpha_q$, whereas the reverse was true for PLC (Rümenapp et al 2001). Since $G\alpha_{13}$ can activate Rho in many cell lines (Buhl et al. 1995; Plonk et al. 1996; Katoh et al. 1998; Hooley et al. 1996; Fromm et al. 1997; Mao et al. 1998; Murthy et al. 2001) and since Rho is an activator of PLD, this seems the probable mechanism by which the M3 muscarinic receptor activates the enzyme. Since the data of Rümenapp et al. (2001) indicate that this receptor also activates PLC via $G\alpha_q$, it is somewhat surprising that PLD is not also activated via a PKC-dependent mechanism. As also shown in earlier reports (Schmidt et al. 1994, 1996, 1998; Voss et al. 1999), PLD can be activated by PMA in these cells. Thus it seems that stable expression of the M3 receptor in the cells may have altered the ability of PLC activation to lead to stimulation of PKC.

Agents that activate Rho family proteins could increase PLD activity by inducing activation and translocation of these proteins to membranes containing PLD1. Several studies have reported rapid agonist-induced membrane translocation of RhoA (Fleming et al. 1996; Kranenburg et al. 1997; Abousalham et al. 1997; Keller et al. 1997; Fensome et al. 1998; Michaely et al. 1999; Houle et al. 1999). It has also been shown that activation of Rho proteins by GTPγS induces their membrane association (Bokoch et al. 1994; Fleming et al. 1996). Although the specific membrane fractions to which the Rho proteins were translocated were not defined in most cases, some studies have shown that these proteins are relocalized to plasma membranes or caveolae (Kranenburg et al. 1997; Michaely et al. 1999). As noted earlier, there is evidence that both PLD1 and PLD2 are present in caveolae.

6.4
Role of ARF Family GTPases

ARF was one of the first in vitro regulators of PLD to be recognized (Brown et al. 1993; Cockcroft et al. 1994). It was recognized to be the principal component of cytosol responsible for PLD activation (Fig. 17, Cockcroft et al. 1994). It exists in six mammalian isoforms (ARF1-6) and all are capable of activating PLD (Massenburg et al. 1994; Brown et al. 1995; Tsai et al. 1998). An ARF-like protein termed hARL1, which is 57% identical in amino acid sequence to hARF1, is also able to activate PLD (Hong et al. 1998). Myristoylation enhances the potency of the ARFs, as seen for most actions of these small G proteins (Massenburg et al. 1994; Brown et al. 1995; Tsai et al. 1998; Fensome et al. 1998).

PLD1 is much more responsive to ARF than is PLD2 (Hammond et al. 1997; Colley et al. 1997; Kodaki & Yamashita 1997; Min et al. 1998; Lopez et al. 1998; Lee et al. 1998; Sung et al. 1999a; Zhang et al. 1999; Slaaby et al.

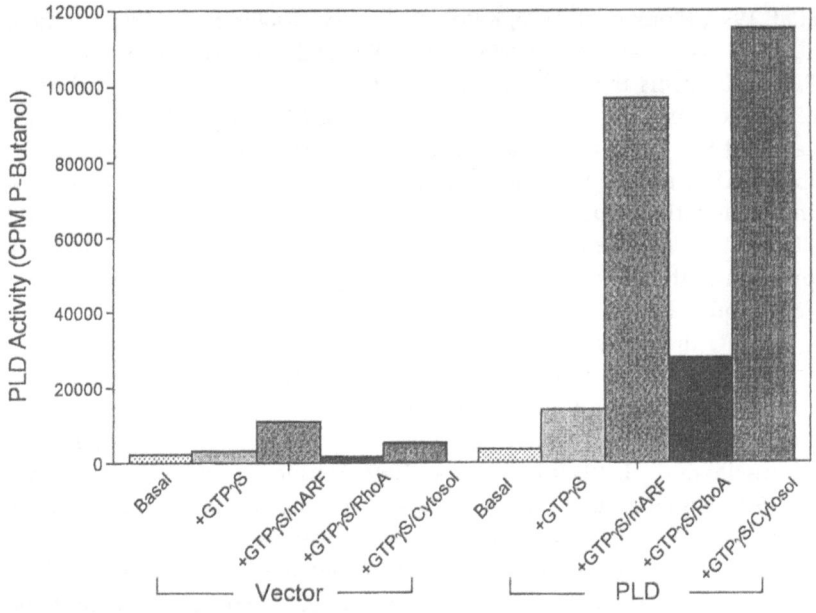

Fig. 17. Effects of ARF, Rho and cytosol on PLD activity of membranes from COS-7 cells expressing vector or rPLD1. Cells were transfected with pcDNA3 vector or with rPLD1 and membranes isolated. PLD activity was measured in vitro by measuring [$^{3+}$H]PtdBut formation from [^{3}H-palmitoyl]PC in PL vesicles. GTPγS was added alone or with myristoylated ARF3, prenylated RhoA or cytosol from the same cells. From Park et al. (1997) by pemission of the authors and publisher

2000). Synergism is observed in the actions of ARF with PKCα or RhoA on the activity of partially purified preparations of PLD (Singer et al. 1996; Hodgkin et al.1999; Kuribara et al. 1995) and on recombinant PLD1 (Hammond et al. 1997).

The interaction site on ARF for PLD has been localized to its N-terminus (Zhang et al. 1994; Jones et al. 1999) including the N-terminal helix and adjacent loop, the α2-helix and part of the β2-strand (Jones et al. 1999). The PLD interaction region is different from that required for activation of cholera toxin (Zhang et al. 1995) or for coatamer binding (Jones et al. 1999). However, the interaction site on PLD1 for ARF has not been defined, although it is not located in the N-terminal 319 amino acids (Park et al. 1998; Sung et al. 1999). As noted above, PLD1 is much more responsive to ARF than is PLD2. However, truncation of the first 308 amino acids of PLD2 reduces its basal activity and renders it very responsive to ARF (Sung et al. 1999).

Although there have been many reports showing that ARF stimulates PLD in membranes and partially purified preparations of the enzyme, and also activates PLD1 in vitro (Brown et al. 1993; 1995; Bourgoin et al. 1995; Singer et al. 1996; Massenburg et al. 1994; Cockcroft 1994; Hammond et al. 1995, 1997; Park et al. 1997; Abousalham et al. 1997; Min et al. 1998b; Sung et al. 1999a, 1999b; Katayama et al. 1998; Rumenapp et al. 1995; Caumont et al. 1998; Hodgkin et al. 1999), the evidence that this occurs in vivo is more limited. Thus expression of constitutively active L^{71}ARF3 in COS7 cells was observed to stimulate the endogenous PLD, but had no effect on expressed PLD1 (Park et al. 1997). This could be due to the localization of expressed PLD1 in the Golgi, which are disrupted by L^{71}ARF3. Better evidence for ARF activation of PLD in vivo has come from studies with ARF6. This class III ARF is very effective in activating PLD (Massenburg et al. 1994) and is principally localized to the plasma membrane (Cavenagh et al. 1996) where it is involved in membrane trafficking and actin remodeling (Al-Awar et al. 2000). ARF6 cycles between plasma membranes and cytosol or a tubulovesicular compartment as determined by its GTP binding/activation state (Gaschet & Hsu, 1999; Radhakrishna & Donaldson, 1997). Stimulation of chromaffin cells with nicotine to elevate the cytosolic Ca^{2+} caused an activation of PLD and also a translocation of ARF6 to the plasma membrane fraction (Caumont et al. 1998) resulting in a concurrent increase in PLD activity in this fraction. Treatment of the permeabilized cells with GTPγS similarly caused an increase in PLD activity and a relocalization of ARF6 to the plasma membrane (Caumont et al. 1998). Further evidence for the control of PLD by ARF6 was obtained in experiments in which the activation of PLD by Ca^{2+} in permeabilized chromaffin cells was markedly inhibited by addition of a myristoylated peptide corresponding to the N-terminus of ARF6, but not by the corresponding peptide from ARF1 (Caumont et al. 1998). Similar results with these peptides were obtained in myometrial extracts treated with GTPγS (Le Stunff et al. 2000a). These results suggest a major role for ARF6 in regulation of PLD in some cell types. The regulation of PLD in the Golgi apparatus by Class I ARFs will be discussed below in Section 7.1.

There have been several reports implicating ARF in the regulation of PLD by certain agonists. In permeabilized HEK cells stably expressing the M3 cholinergic receptor, the activation of PLD by GTPγS was reduced by cytosol depletion and restored by ARF1 (Rümenapp et al. 1995, 1997). The activation of PLD by carbachol in these cells was inhibited by brefeldin A (BFA) (Rümenapp et al, 1995), which is a fungal metabolite inhibitor of some of the guanine nucleotide exchange factors (GEFs) for ARFs (Moss & Vaughan, 1998). Another study showed that BFA (5–50 μg/ml) inhibited PDGF and PMA activation of PLD in Rat1 fibroblasts overexpressing insulin receptors (HIRcB cells) whereas an inactive analog was without effect (Shome et al. 1998). Similarly, BFA has been reported to inhibit the effects of angiotensin II, endothelin 1 and PDGF on PLD in vascular smooth muscle cells (Shome et al. 1999; Andresen et al. 2001) and the actions of histamine, bradykinin and carbachol on the enzyme in 1321N1 astrocytoma cells (Mitchell et al. 1998). Activation of M3 muscarinic cholinergic or AT_1 angiotensin II receptors in the latter cells resulted in their co-immunoprecipitation with ARF1 and RhoA.. Co-immunoprecipitation and BFA inhibition were not observed with activation of receptors for gonadotropin-releasing hormone (GnRH) (Mitchell et al. 1998). However, these attributes could be restored by mutation of these receptors to contain the canonical Asp Pro XX Tyr sequence present in the seventh transmembrane domain in the effective receptors. Thus this motif was proposed to mediate interaction of the receptors with RhoA and ARF and hence activation of PLD (Mitchell et al. 1998).

In contract to the preceding findings, BFA has been reported to be without effect on basal PLD in A549 adenocarcinoma cells or on the stimulation of the enzyme by PMA, sphingosine1-phosphate or bradykinin (Meacci et al. 1999). In HL-60 cells, the inhibitor also did not inhibit the stimulation of PLD by formyl-Met-Leu-Phe (FMLP) or ATP at times and doses when the Golgi apparatus was completely disrupted (Guillemain & Exton, 1997). These data suggest that, if ARF is involved in agonist activation of PLD in these cells, its activation does not depend on a BFA-sensitive GEF.

There have been other reports implicating ARF in agonist activation of PLD. For example, in permeabilized HIRcB cells that overexpress insulin receptors, insulin has been reported to enhance the effect of ARF plus GTPγS on PLD (Shome et al. 1997). Insulin was also reported to promote the binding of GTPγS to ARF and the binding of ARF to cell membranes. In another study, expression of dominant negative forms of ARF1 and ARF6 in HIRcB cells resulted in inhibition of the stimulation of PLD by PDGF and PMA (Shome et al.1998). FMLP and PMA have also been shown to promote the association of ARF to membranes in intact HL-60 cells (Houle et al. 1995, 1999), which was correlated with an increase in GTPγS-stimulated PLD activity in the membranes.

In cytosol-depleted neutrophils stimulated with FMLP, myristoylated ARF1 restored the stimulation of PLD by GTPγS (Fensome et al. 1998). Pertussis toxin blocked the effect of FMLP implying the involvement of G proteins of the Gi/Go family. The C3 exoenzyme of *C. botulinum* partly reduced the effect, suggesting synergism with Rho family proteins (Fensome et al. 1998). FMLP, PMA and GTPγS also promoted the membrane translocation of both ARF and Rho in the intact or permeabilized neutrophils. In another study, subcellular fractionation of HL-60 cells revealed that ARF1-dependent and FMLP-dependent PLD activity was located in plasma membranes and endomembranes (Whatmore et al. 1996). Substantial activity of PI-4 kinase and PI 4P 5-kinase was found in plasma membranes. PIP2 was enriched in these membranes, whereas PI and PI 4-P were found predominantly in endomembranes. These findings support roles for ARF and PIP_2 in the regulation of PLD in plasma membranes in these cells.

An additional way in which ARFs could regulate PLD activity is through the effects of ARF1 and ARF6 on PI-4P 5-kinase, which synthesizes PIP_2. ARF1 directly interacts with Type I PI-4P 5-kinase and increases its activity (Jones et al. 2000). Addition of ARF1 to Golgi membranes in the presence of the type I enzyme also dramatically increases the PIP_2 level in these membranes (Jones et al. 2000). Another report has shown that ARF can recruit PI 4-P 5-kinase-β to the Golgi complex, resulting in a potent stimulation of PIP_2 synthesis (Godi et al. 1999). In both studies, the effects were independent of PLD activity. Another report has identified PI 4-P 5-kinase as a downstream effector of ARF6 (Honda et al. 1999) in brain cytosol and HeLa cells. These proteins were colocalized in membranes ruffles in cells activated with AlF_4^- or EGF. In contrast to ARF6, neither ARF1 nor ARF5 was colocalized with the lipid kinase, whereas PLD2 relocalized to the ruffles perhaps because of a local increase in PIP_2 (Honda et al. 1999).

Another approach to exploring the role of ARF in the activation of PLD in vivo has utilized proteins that inhibit ARF action or activation. Arfaptin an ARF binding protein (Kanoh et al. 1997) inhibited the activation of PLD by different ARFs in vitro (Fig. 18, Tsai et al. 1998; Williger et al. 1999b) and by PMA in NIH 3T3 cells (Williger et al. 1999a). Similarly, ARP an ARF-related protein inhibited the in vitro stimulation of PLD by ARF and ARNO, a GEF for ARF (Schürmann et al. 1999). The constitutively active form of ARP was

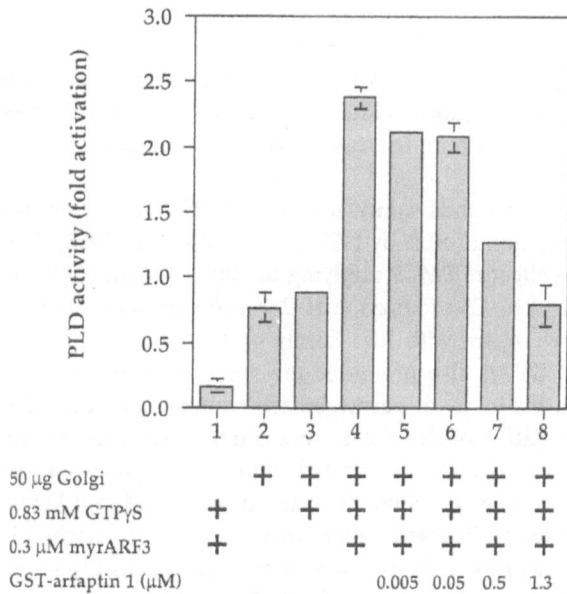

Fig. 18. Arfaptin inhibits ARF stimulation of PLD activity in Golgi. Golgi-enriched membranes were incubated with GTP(S, myristoylated ARF3 and increasing concentrations of arfaptin. PLD activity was measured by [³H]PtdBut formation from [³H-palmitoyl]PC in PL vesicles. From Williger et al. (1999) by permission of the authors and publisher

also inhibitory, but the dominant negative form was not, indicating that ARP was acting by interacting with ARNO. The active and wild-type forms of ARP also inhibited muscarinic activation of PLD and ARF translocation to membranes (Schürmann et al. 1999). These data suggest that ARP inhibits ARF-mediated PLD activation by binding to ARNO or other proteins containing the Sec7 (GEF) domain.

6.5
Role of Tyrosine Phosphorylation

Many agonists whose receptors encode tyrosine kinase activity or which activate soluble tyrosine kinases can stimulate PLD (Natarajan et al. 1996; Exton 1997, 1999). Thus growth factors acting on many cell types, and agents that activate T cell receptors or the receptors for IgE on mast cells increase the activity of the enzyme. However, the activation mechanism is indirect in most cases and may not involve phosphorylation of the enzyme. Growth factors can activate multiple signaling pathways leading to regulation of the enzyme by PKC, Rho family proteins and Ras family proteins (Exton 1997).

The use of inhibitors of tyrosine kinases has also implicated these kinases in the activation of PLD by agonists acting through heptahelical receptors that couple to heterotrimeric G proteins. Thus, the inhibitors decrease the activation of PLD by ATP in U937 promonocytes (Dubyak et al. 1993; Kusner et al. 1993), thrombin in platelets (Martinson et al. 1994), FMLP in neutrophils or HL60 cells (Uings et al. 1992; Houle et al. 1999), endothelin 1, norepinephrine and angiotensin II in vascular smooth muscle cells (Wilkes et al. 1993; Jinsi et al. 1996; Suzuki et al. 1996b), bombesin in fibroblasts (Briscoe et al. 1995), carbachol in PC12 pheochromocytoma cells and HEK cells expressing the M3 muscarinic receptor (Ito et al. 1997a; Schmidt et al. 1994) and endothelin 1 in myometrium (Le Stunff et al. 2000b).

Another approach to identifying a role for tyrosine phosphorylation in the regulation of PLD involves the use of vanadate, which is an inhibitor of tyrosine phosphatases. Stimulation of PLD activity by vanadate or H_2O_2 alone or in combination has been demonstrated in HL-60 cells (Bourgoin & Grinstein 1992), U937 cells (Dubyak et al. 1993), endothelial cells (Natarajan et al. 1993), vascular smooth muscle (Ward et al. 1995), Swiss 3T3 fibroblasts

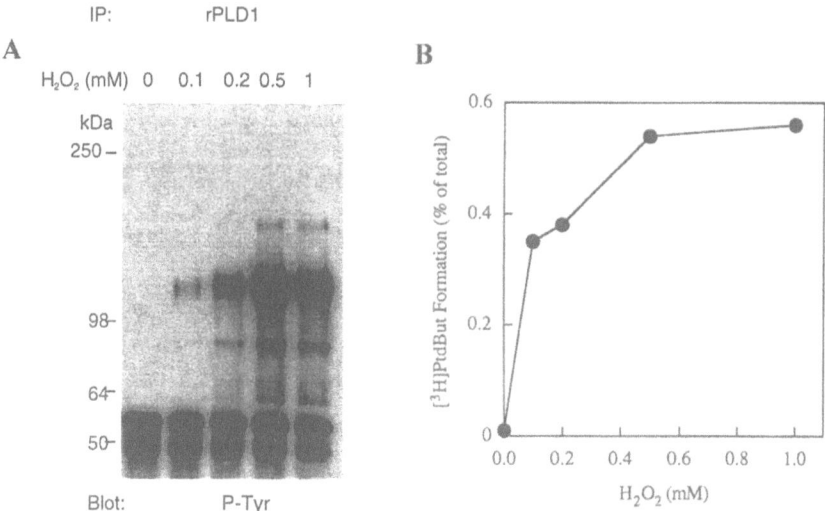

Fig. 19A, B. H_2O_2 stimulation of PLD activity and protein tyrosine phosphorylation in Swiss 3T3 fibroblasts. The cells were labeled with [³H]myristic acid, pretreated with vanadate and then stimulated with H_2O_2 at different concentrations. PLD was measured by [³H]PtdBut formation, and all lysates were immunoprecipitated using an anti-rPLD1 antibody then Western blotted with an anti-PTyr antibody. Adapted from Min et al. (1998a) by permission of the authors and publisher

(Min et al. 1998a) and myometrium (Le Stunff et al. 2000). A key point in the regulation of PLD activity by tyrosine kinases is whether or not the activity of the enzyme is controlled by direct tyrosine phosphorylation. On this point there is no clear evidence. Several studies have shown that H_2O_2 activates PLD and promotes the tyrosine phosphorylation of several cellular proteins including PLD (Fig. 19, Gomez-Cambronero 1995; Ito et al. 1997b; Marcil et al 1997; Min et al. 1998a). Although the tyrosine phosphorylation of cellular proteins was correlated with the activation of PLD, the tyrosine phosphorylation of PLD itself was not (Bourgoin & Grinstein 1992; Ito et al. 1997a, 1997b; Min et al. 1998a). Furthermore, the effects of PKC inhibitors and PKC down-regulation indicated that PKC mediated most of the PLD activity change seen with H_2O_2 in Swiss 3T3 cells (Min et al. 1998a) although not in endothelial cells (Natarajan et al. 1993). Another study showed that mutation of the tyrosine phosphorylation site in PLD2 did not alter the ability of EGF to activate the enzyme in HEK 293 cells expressing both PLD2 and the EGF receptor (Slaaby et al. 1998). In contrast to the marked tyrosine phosphorylation of PLD induced by H_2O_2 in Swiss 3T3 cells, that induced by PDGF was barely detectable, yet the growth factor induced a large activation of the enzyme (Min et al. 1998a).

A recent report showed that norepinephrine increased the tyrosine phosphorylation of PLD2 in vascular smooth muscle cells (Parmentier et al. 2001). The effect was blocked by U0126 a reported inhibitor of MAP kinase kinase. However, ERK2 a MAP kinase did not phosphorylate PLD2 invitro, and the kinase responsible for the in vivo phosphorylation and the effect of this phosphorylation on PLD2 activity remain unknown.

The nature of the tyrosine kinases activated by G protein-coupled receptors and H_2O_2 is not well defined, but there is evidence that they are Ca^{2+}-dependent (Ito et al. 1997a, 1997b). All of the G protein-coupled receptors that mediate PLD activation through a tyrosine kinase mechanism are known to activate G_q and phospholipase C, and thus increase cytosolic Ca^{2+} (Exton 1996). This would then activate any Ca^{2+}-dependent tyrosine kinases. In fact, angiotensin II has been shown to increase tyrosine kinase activity in a Ca^{2+}-dependent manner in liver epithelial cells and aortic smooth muscle cells (Huckle et al. 1992) and this is true for thrombin acting on BC_3H1 muscle cells (Offermanns et al. 1993) and platelet-activating factor acting on Kupffer cells (Chao et al. 1992), although with other agonists and cell lines, there is evidence that other, Ca^{2+}-independent protein kinases are involved. A focal adhesion kinase homologue termed PYK2 (also known as CADTK, CAKβ, RAFTK and FAK2) has been implicated in the effects of LPA, angiotensin II, cholecystokinin and PDGF in several cell lines (Yu et al. 1996; Brinson et al. 1998; Tapia et al. 1999; Tang et al. 2000), whereas focal adhesion kinase itself seems not to be involved in Ca^{2+}-mediated tyrosine phosphorylation (Sinnett-Smith et al. 1993). In the case of H_2O_2, there is evidence that Syk is the tyrosine kinase specifically activated in a B cell line (Schieven et al. 1993), whereas in a T cell line, Lck, another Src family member is involved (Hardwick & Sefton 1997).

Another mechanism by which agonists acting through G protein-coupled receptors could act via tyrosine phosphorylation is through transactivation of the EGF receptor (Prenzel et al. 1999; Zwick et al. 1999; Keely et al. 2000). This is thought to involve Ca^{2+} activation of a metalloprotease that cleaves a membrane precursor to release HB-EGF which then acts on the EGF receptor (Prenzel et al. 1999). This mechanism has not yet been examined for the activation of PLD by G protein-coupled receptors.

6.6
Roles of Ras and Ral

Several reports have indicated that tyrosine phosphorylation or G protein-mediated signaling can regulate PLD through the low Mr GTPases Ras and Ral in some cell lines. Thus, infection of BALB/c 3T3 cells with v-Src followed by activation of its tyrosine kinase activity resulted in stimulation of PLD (Song et al. 1991) by a mechanism that was mediated by Ras (Jiang et al. 1995a), but required additional cytosolic factors. In further work, one of these factors was shown to be Ral, a member of the Ras subfamily, and this GTPase was shown to associate with PLD, but not to activate it (Jiang et al. 1995b). Thus it was proposed that Ras interacted with a GEF for Ral leading to Ral activation which, together with another factor(s), stimulated the enzyme.

Other reports showed a functional association between RalA and PLD1 in cell lysates, but this alone did not activate the enzyme (Jiang et al. 1995b; Luo et al. 1997, 1998). Further work demonstrated that ARF could also associate with active RalA-PLD1 complexes isolated from v-Src-transformed cells, and the association was increased when ARF was in the active, GTP-bound form (Luo et al. 1998). Another group found that ARF1 and RalA could bind to PLD1 and that the interaction sites were different (Kim et al. 1998). Furthermore, RalA was observed to enhance the effect of ARF1 on the activity of PLD1 (Fig. 20, Kim et al. 1998).

Another approach to exploring the role of Ral in the regulation of PLD involves the use of the lethal toxin (TcsL) from C. Sordellii and a variant of another toxin (TcdB-1470) from C. difficile. These toxins inactivate certain low Mr GTPases by monoglucosylation, includingg Rac, Ras, Ral and Rap (Popoff et al. 1996; Schmidt et al. 1998). Long-term treatment of HEK-293

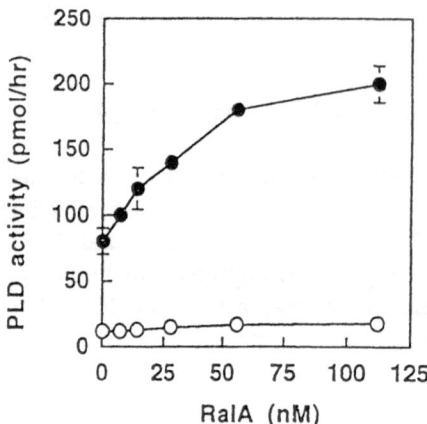

Fig. 20. RalA enhances the stimulatory effect of ARF on PLD activity. Increasing concentrations of GTPγS-activated RalA were added to incubations of purified PLD1. PLD activity was measured in vitro by the release of [³H]choline from [³H]choline-labeled PC. From Kim et al. (1998) by permission of the authors and publisher

cells with TcsL or TcdB-1470 caused inhibition of PLD activation by PMA, but did not affect the basal activity of the enzyme or the effects of carbachol or GTPγS, when the latter was added to the permeabilized cells (Schmidt et al. 1998). The addition of Rac1, Ras or Rap did not restore the PMA-stimulated activity, but it was fully restored by RalA or RalB. These results suggest that Ral is involved in PKC stimulation of PLD in HEK-293 cells. Another group found that overnight treatment with TcsL inhibited the effect of GTP(S to stimulate PLD in permeabilized HL-60 cells (El Hadj et al. 1999). It also reduced the stimulatory effects of ARF, Rho and Rac in vitro and of PMA in intact cells. Since ARF and Rho are not targets of the toxin, these results may have been due in part to the reduction in PIP2 induced by the toxin (El Hadj et al. 1999). Experiments with TcsL toxins that were active or inactive on Ral indicated that this GTPase was involved in the stimulation of PLD by GTPγS.

As noted above, there is evidence that constitutively active Ras activates PLD in intact cells (Jiang et al. 1995a; Carnero 1994a, 1994b; Price et al.1989b; Lucas et al. 2000). On the other hand, Ras does not directly activate PLD in vitro (Fig. 15, Quilliam et al. 1990; Hurst et al. 1991; Bae et al. 1998). Therefore the effect of Ras on PLD in vivo must involve a signaling path-way(s). As discussed above, there is evidence that one pathway involves Ral. The involvement of Ras and Ral in the regulation of PLD by growth factors and other agonists signaling through tyrosine phosphorylation has been indicated in several studies. Thus expression of dominant negative forms of Ras and Ral inhibited the activation of PLD by epidermal growth factor (EGF) in Rat 3Y1 cells expressing the EGF receptor (Lu et al. 2000). Overexpression of wild type or activated RalA increased basal PLD activity and transformed the cells, and transformation was also observed in cells overexpressing PLD1 (Lu et al. 2000). In another report, expression of dominant negative Ras in NIH 3T3 cells completely blocked the activation of PLD by PDGF whereas expression wild type Ras in Rat 2 cells enhanced the response (Lucas et al. 2000). On the other hand these manipulations had minimal effects on the activation of PLD by phorbol ester. Expression of the adaptor molecules Shc and Grb2 in Rat 2 cells also amplified the effect of PDGF on PLD. These results support the operation of another, Ras-dependent, pathway for growth factor activation of PLD. A role for Ras and Ral in PLD activation by G protein-coupled receptors has also been indicated. In HEK-293 cells expressing the M3 muscarinic receptor, PMA-, but not carbachol-induced activation of PLD was reduced by expression of dominant negative forms of RalA and Ras, and overexpression of the Ral-specific GEF Ral-GDS enhanced PKC-induced PLD stimulation (Voss et al. 1999). The stimulatory effect of Ral-GDS was abolished by toxin TcdB-1470 from *C. difficile*, and the toxin and also PKC inhibition blocked the stimulation of PLD by EGF and PDGF (Voss et al. 1999). However the effects of carbachol were not modified by TcdB-1470, PKC inhibition or by expression of Ral-GDS. These data indicate that, in this cell type, PKC and Ral mediate the effects of PMA and growth factors on PLD, but that muscarinic stimulation is not dependent on PKC or Ral. Since Ral is activated by PMA or growth factors (Voss et al. 1999), it seems that this Low Mr GTPase is downstream from PKC in the signaling pathway. Other studies have implicated the Ras-Raf-MAP kinase pathway in the activation of PLD. One study found that PLD activity was elevated in v-Raf transformed cells and used dominant negative forms of Ral and Rho to indicate that these GTPases were downstream from v-Raf (Frankel et al. 1999). Another used inhibitors to implicate Ras and MAP kinase in the activation of PLD by norepinephrine (Muthalif et al. 2000). More work is needed to substantiate the role of the Ras-Raf-MAP kinase pathway in PLD regulation.

6.7
Roles of Calcium and Calmodulin

Treatment of several cell types with Ca^{2+} ionophores can activate PLD, and Ca^{2+} chelators can inhibit the stimulation of PLD by several agonists (Exton 1997). However studies of the effects of Ca^{2+} and Mg^{2+} on recombinant or purified PLD show that the enzyme is stimulated by submicromolar concentrations of Ca^{2+}, but that the stimulation is not seen in the presence of physiological concentrations of Mg^{2+} (Brown et al. 1995; Hammond et al. 1997; Min et al. 1998). These latter observations render it unlikely that Ca^{2+} *directly* activates the enzyme under physiological conditions, and indicate that Ca^{2+} stimulation of the enzyme in vivo involves other factors. Examination of the Ca^{2+}-dependence of the effects of PMA, vasopressin or FMLP on PLD activity in neutrophils or hepatocytes, and of GTPγS and ARF effects in permeabilized neutrophils have indicated that increasing the cytosolic Ca^{2+} concentration from 0.1 μM to 0.5 μM or higher enhances basal PLD activity and also the effects of the various agents on the enzyme (Kessels et al. 1991; Gustavsson et al. 1994; Cockcroft et al. 1994). These data indicate that variations in cytosolic Ca^{2+} within the physiological range can influence the regulation of PLD.

Likely mediators of the effects of Ca^{2+} on PLD are the Ca^{2+}-dependent isozymes of PKC. This appears to be the case in hepatocytes, where PKC inhibitors reduce the stimulatory effect of Ca^{2+} (Gustavsson et al. 1994). As noted above in the section (6.2) on PKC, there is much evidence that Ca^{2+}-dependent α- and β-isozymes are the major PKC isoforms that regulate PLD. Another possible mediator is calmodulin (CaM). Although PKCα and PKCβ play some role in the activation of PLD by antigen in RBL-2H3 basophilic cells, the effects of CaM antagonists indicate that this Ca^{2+}-binding protein also plays a role (Kumada et al. 1995). A similar situation is seen in studies with CHO cells expressing muscarinic receptors, where PLD activation by carbachol is decreased by PKC down-regulation and also by Ca^{2+} chelation and inhibition of Ca^{2+}/CaM-dependent protein kinase II (Min et al. 2000).

In related studies, overexpression of a CaM binding myristoylated alanine-rich PKC substrate termed MARCKS in SK-N-MC neuroblastoma cells has been observed to increase PLD stimulation by PMA (Morash et al. 1998). PKCα, but not PKCβ, was up-regulated in the cells and, like MARCKS, PKCα was lost when the cells were at a high passage number. The mechanism of the MARCKS effect is unknown, but it may act by colocalizing PKCα and PLD and/or sequestering PIP_2 (Morash et al. 1998). In further studies, MARCKS overexpression was found not to affect GTPγS-stimulated PLD activity in the permeabilized cells (Morash et al. 2000), and the GTPγS effect on PLD was independent of PKC. Western blot analysis of a detergent-insoluble fraction of the cells showed the presence of MARCKS, PKCα, PLD1 and PLD2, suggesting the colocalization of these proteins in caveolae.

6.8
Regulation by Other Proteins

Some early reports of the regulation of PLD by ARF indicated that other cytosolic factors were required (Houle et al. 1995; Bourgoin et al. 1995; Singer et al. 1995). One of these factors was identified as PKC((Singer et al. 1996) which is consistent with the ability of this enzyme to synergize with ARF on PLD1, as discussed in preceding sections. Protein inhibitors of the enzyme have also been described. Fodrin, a homolog of spectrin, was shown to inhibit PLD activity (Lukowski et al. 1996) and the mechanism was later ascribed to a decrease in PIP_2 (Lukowski et al. 1998). Likewise, another inhibitor of PLD (Han et al. 1996) turned out to be synaptojanin, a phosphatase that hydrolyzes PIP_2 (Chung et al. 1997). A PLD inhibitor that apparently does not act by altering PIP_2 is the clathrin assembly protein 3, which interacts with PLD directly (Lee et al. 1997). PLD2 has high intrinsic activity and it has been postulated that natural inhibitors of this enzyme exist (Colley et al. 1997). Possible inhibitors are α- and β-synucleins since these brain proteins selectively inhibit PLD2 in comparison to PLD1 (Jenco et al. 1998). Myocardial PLD2 associates with α-actinin, which exerts an inhibitory effect (Park et al. 2000). α-Actinin interacts directly with the N-terminus of PLD2 and this interaction and its inhibitory action are reversed by ARF1, which releases α-actinin from its binding site.

β-Actin itself has recently been reported to bind directly to PLD1 and PLD2 and to inhibit their catalytic activity when stimulated by PIP_2, oleate or ARF1 (Lee et al. 2001). The binding site was localized to the 613–723 amino acid sequence of PLD2. β-Actin could displace α-actinin binding to PLD2, but had minimal effect on the ability of ARF1 to activate the enzyme. Immunoprecipitation studies demonstrated that actin interacted with both PLD1 and PLD2 in PC12 or COS7cells. Furthermore, immunocytochemical studies indicated that the two isozymes co-localized with F-actin (Lee et al. 2001). These findings indicate that actin can interact with PLD1 and PLD2 in vivo and may exert an inhibitory effect on both enzymes in intact cells.

6.9
Regulation by Ceramide

Ceramide is a product of the breakdown of sphingomyelin by sphingomyelinase, but can also be synthesized de novo or produced from sphingosine or ceramide 1-phosphate (Hannun 1996; Spiegel et al. 1996). It participates in cell signaling and plays roles in stress responses such as apoptosis, cell cycle

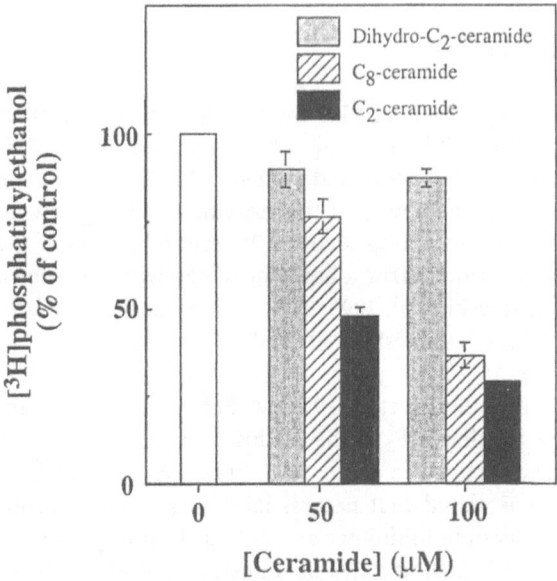

Fig. 21. Inhibition of PLD activity by ceramide. Membranes and cytosol from [^{3}H]myristate-laabeled HL60 cells were incubated with ethanol and GTPγS in the presence or absence of C$_2$-ceramide, C$_8$-ceramide or dihydro-C$_2$-ceramide. PLD was assayed in vitro by measuring [^{3}H]ethanol formation. From Abousalham et al. (1997) by permission of the authors and publisher

arrest and inflammation. A large number of agonists can increase sphingo-myelinase activity leading to increased levels of ceramide (Levade and Jaf-frézou 1999). Cell-permeable (C$_2$- or C$_6$-) ceramides produce a marked in-hibition of PLD activity in intact fibroblasts stimulated with PMA (Gómez-Muñoz et al. 1994) and in neutrophils stimulated with FMLP (Nakamura et al. 1994). They also inhibit the ability of exogenous PA or PLD from S. chromofuscus to stimulate DNA synthesis, and reduce the stimulation of PLD exerted by GTPγS or PMA in permeabilized fibroblasts. In a further study, the ceramides were shown to block the activation of PLD by sphingos-ine 1-phosphate (Gómez-Muñoz et al. 1995). Another group reported that treatment of cells with sphingomyelinase or ceramide inhibited PLD activa-tion induced by DOG (dioctanoylglycerol) or bradykinin, but felt that this could be attributed to the inhibition of PKC(translocation (Jones & Murray 1995). Ceramide has also been shown to inhibit IgE-mediated PKC translo-cation and PLD activation in RBL basophilic cells (Nakamura et al. 1994) and to inhibit the stimulation of the enzyme by FMLP in intact HL-60 cells or by GTP(S in permeabilized cells (Fig. 21, Abousalham et al. 1997). As in the report of Jones and Murray (1995), ceramide blocked the translocation of the conventional (Ca^{2+}-dependent) PKC isozymes, but not that of the δ-, ϵ- or ζ-isozymes. These data suggest that in these cell lines, conventional, but not Ca^{2+}-independent or atypical PKC isozymes, are involved in regulation of PLD. In the report of Abousalham et al. (1997) ceramide was also noted to inhibit the membrane association of ARF1, RhoA and Cdc42 induced by FMLP. This could partly explain the impairment of PLD activation. However, a recent report has shown that C_2-ceramide, but not C_8- or dihydro-C_2-ceramide, can directly inhibit purified recombinant PLD1 and block activation of the enzyme in permeabilized chromaffin cells (Vitale et al. 2001). Thus direct effects of ceramides are probably involved.

6.10
Role of Phosphoinositide 3-Kinase

PIP_3 can activate PLD1 directly, although it is not as efficacious as PIP_2 (Hammond et al.1997; Min et al. 1998b). However, the cellular level of PIP_2 is so much higher than that of PIP_3 that variations in the concentration of PIP_3 per se would seem unlikely to influence PLD activity in vivo. Thus the PI 3-kinase inhibitor wortmannin has been reported to be without effect on basal and GTPγS-stimulated PLD activity in permeabilized cells (El Hadj et al. 1999).

Studies of the regulation of PLD by PI 3-kinase in intact cells have principally utilized wortmannin and LY294002, and have given variable results. Thus, wortmannin is without effect on the stimulation of PLD by PDGF or EGF in Rat1 fibroblasts (Hess et al. 1997). Neither does it or LY294002 inhibit the activation of the enzyme by angiotensin II in vascular smooth muscle cells (Andresen et al. 2001). On the other hand, the inhibitors block the stimulation of PLD by the high affinity IgG receptor FcγRI in U937 cells (Gillooly et al. 1999) and the activation of the enzyme in endothelial cells by stem cell factor (Kozawa et al. 1997). Using a different approach, namely using cells expressing wild type or mutant PDGF receptors, no evidence could be obtained for a role of PI 3-kinase in PDGF activation of PLD (Yeo et al. 1994). An obvious explanation for these divergent results is that different cells and agonists use different pathways to activate PLD and any role for PIP3 probably lies upstream from PLD.

7
Cellular Functions of Phospholipase D

Despite its widespread distribution and its regulation by many agonists, the cellular functions of PLD remain ill-defined. As described below, the effects of exogenous PLD or PA added to cells are complicated by the generation of other signaling molecules besides PA. Approaches using stable overexpression of PLD isozymes have, with a few exceptions, led to cell death, and deletion of these enzymes by antisense methods has not been very successful. There have been no reports of the knockout of either PLD1 or PLD2 by homologous recombination in mice. On the other hand, there have been a few reports of the use of catalytically inactive PLD isozymes as dominant negative alleles to reduce PLD activity in some cells. Studies of the role of PLD in specific cell functions are described below.

7.1
Role in Vesicle Trafficking in Golgi

There is much evidence that PLD and PA play roles in vesicle trafficking in the Golgi, but even this is disputed. Class I ARFs are well known to be involved in trafficking in Golgi through their roles in the recruitment of COP proteins and adaptor proteins (APs) to Golgi membranes, which results in the formation of COP-I-coated or clathrin-coated vesicles (Donaldson & Lippincott-Schwartz, 2000; Stamnes & Rothman 1993; Traub et al. 1993). ARF-stimulated PLD activity has also been reported in Golgi (Provost et al.1996; Liscovitch 1999; Ktistakis et al. 1995; Chen et al. 1997) and most immunocytochemical studies have localized PLD1 to the perinuclear region, which contains this organelle (Colley et al. 1997; Sugars et al. 1999; Kam & Exton, 2001; Freyberg et al. 2001).

The presence of ARF-stimulated PLD activity in the Golgi has given rise to the proposal that this enzyme plays a role in vesicle trafficking in this organelle (Kahn et al. 1993). In support of this, it was observed, that in a cell line with high constitutive PLD activity, ARF was not required for the formation of coatomer (COP-I)-coated vesicles (Ktistakis et al. 1996). Furthermore, ethanol inhibited the formation of coated vesicles in Golgi incubated with ARF and GTPγS and this was attributed to inhibition of PA formation by PLD. Addition of bacterial PLD to Golgi membranes also induced coatomer binding and coated vesicle formation in the absence of ARF (Ktistakis et al. 1996). Finally, coatomer was observed to bind more avidly to vesicles containing PIP_2 and PA. In an extension of this work, primary alcohols were found to inhibit the transport of viral glycoproteins from the endoplasmic reticulum (ER) to the Golgi, and the block was reversed by liposomes containing PA (Bi et al. 1997).

Another group has provided evidence that PLD plays a role in the release of secretory vesicles from the trans-Golgi network. Using permeabilized GH3 pituitary cells, immunoaffinity-purified PLD1 was observed to stimulate nascent secretory vesicle budding from the trans-Golgi network, whereas 1-butanol inhibited this process (Fig. 22, Chen et al. 1997). Furthermore, generation of PA by a combination of PLC and DAG kinase also stimulated vesicle budding, and PA was found to accumulate in the Golgi during this process (Siddhanta & Shields1998). Primary butanol was also reported to disrupt the structural organization of the Golgi and to inhibit the sorting of polypeptide hormones into post-Golgi vesicles (Siddhanta et al. 2000). It was shown that inhibition of PA formation resulted in decreased PIP2 synthesis and it was suggested that this lipid was required for maintaining the structural integrity and function of the Golgi. Other work has indicated a requirement for PIP_2 in secretory vesicle formation (Fensome et al.1996; Tüscher et al. 1997; Way et al. 2000).

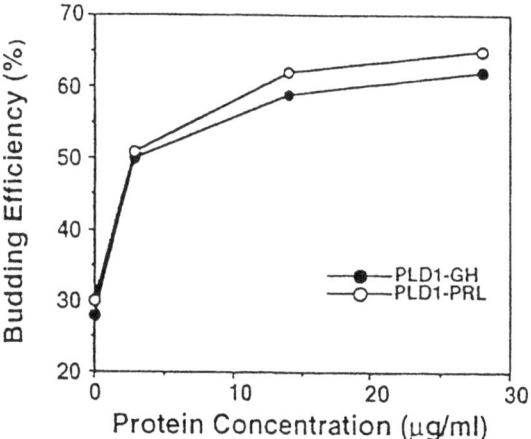

Fig. 22. PLD stimulates the release of nascent secretory vesicles from the trans-Golgi network. GH3 pituitary cells were labeled with an amino acid mixture containing [^{35}S]methionine and permeabilized. The cells were then incubated with Sf9 cell lysates expressing hPLD1, and the supernatant and pellet fractions immunprecipitated with antibodies to growth hormone (GH) or prolactin (PRL). Vesicle budding efficiency was calculated as the immunoreactive material in the supernatant divided by that in the supernatant plus pellet. The protein concentrations refer to that of the hPLD1 preparations used. From Chen et al. (1997) by permission of the authors and publisher

In contrast to these reports, Stamnes et al. (1998) measured PA levels in Golgi membranes and found that they declined, rather than increased, during cell-free budding of coatomer-coated vesicles. Another group (Kuai et al. 2000) examined the effects of mutations in ARF3 on its ability to activate PLD1 and to recruit coatomer to Golgi membranes. They observed a very poor correlation between these effects, implying that PLD activation was not critical for Golgi vesiculation. Similarly, Jones et al. (1999) reported that an N-terminally deleted dominant negative mutant of ARF1 that did not inhibit ARF1-stimulated PLD activity effectively competed with ARF1 to prevent coatomer binding to Golgi membranes. Another study of the recruitment of AP-1 adaptors to trans-Golgi membranes reported that this was not affected by exogenous PLD or by neomycin, which is a high-affinity ligand of PIP_2 (West et al. 1997). Thus a role for PLD in Golgi trafficking remains controversial.

7.2
Role in Exocytosis and Endocytosis

PLD has also been implicated in exocytosis and endocytosis. Thus, secretion in mast cells, mammary epithelial cells, platelets, neutrophils and HL60 cells is inhibited by the addition of primary alcohols (Fig. 23, Stutchfield & Cockcroft 1993; Gruchalla et al. 1990; Lin et al. 1991; Benistant & Rubin 1990;Yuli et al. 1982; Fensome et al. 1996; Williger et al. 1999; Brown et al. 1998; Siddhanta et al. 2000; Way et al. 2000; Boisgard & Chanat 2000). However, this could be due to changes other than inhibition of PA formation by PLD. Further support for a role of PLD in exocytosis has come from several approaches. Williger et al. (1999) observed that secretion of metalloproteinase-9 from HT1080 fibrosarcoma cells could be induced by PA, but not by DAG. Subcellular fractionation studies have identified PLD activity in secretory granules in neutrophils and HL60 cells (Whatmore et al. 1996; Morgan et al. 1997) and stimulation of these cells by FMLP causes translocation of the PLD-containing vesicles to the plasma membrane. Likewise, in RBL-2H3 mast cells expressing GFP-tagged PLD1, the translocation of the enzyme from secretory granules to the plasma membrane was stimulated by cross-linking the IgE receptor (Brown et al. 1998).

The final stage in exocytosis is fusion of secretory vesicles with the plasma membrane and release of their contents into the extracellular medium. There is evidence that PLD is involved in this process. Thus in chromaffin cells, stimulation or elevation of cytosolic Ca^{2+} leads to a rapid translocation of ARF6 from secretory granules to the plasma membrane and a concomitant activation of PLD in this membrane (Caumont et al. 1998).

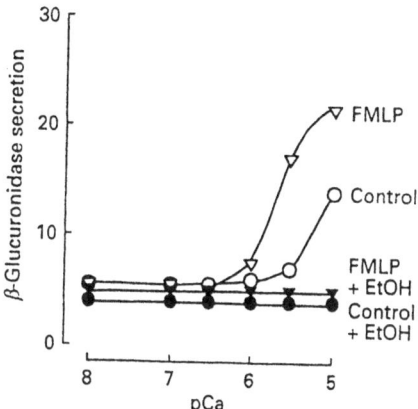

Fig. 23. Ethanol inhibits β-glucuronidase secretion induced by FMLP. Metabolically inhibited HL60 cells were permeabilized with streptolysin O and incubated with and without ethanol and FMLP in the presence of different concentrations of free Ca^{2+}. From Stutchfield & Cockcroft (1993) by permission of the authors and publisher

Furthermore, a myristoylated peptide corresponding to the N-terminus of ARF6 inhibited both PLD activity and the secretion of catecholamine. In later work, antibodies to ARNO, a BFA-insensitive guanine nucleotide exchange factor for ARF, were found to inhibit both PLD activity and catecholamine secretion (Caumont et al. 2000). Expression of an inactive ARNO mutant also blocked secretion of expressed growth hormone in PC12 cells. These studies implicate both ARF6 and PLD in exocytosis at the plasma membrane.

More recent work has shown a close correlation between PLD activation and catecholamine secretion in chromaffin cells stimulated with nicotine and treated with various ceramides (Vitale et al. 2001). Overexpression of wild type PLD1, but not PLD2, in PC12 cells increased exocytosis, whereas expression of catalytically inactive PLD1 inhibited this process. Inactive PLD1 also inhibited nicotine-stimulated catecholamine secretion from chromaffin cells (Vitale et al. 2001). Mutant forms of PLD that did not respond to PKC or did not interact with PIP_2 showed impaired ability to enhance exocytosis compared with wild type PLD (Vitale et al. 2001). These data provide support for a role of PLD1 in the exocytotic process and indicate supporting roles for PKC and PIP_2.

There is also evidence for a role of PLD in endocytosis. In response to stimulation, membrane receptors are endocytosed and degraded. Shen et al. (2001) have reported that the internalization and degradation of EGF receptors in rat fibroblasts is inhibited by primary, but not secondary alcohols.

Furthermore, overexpression of PLD1 or PLD2 resulted in reduced EGF receptor levels, but this was not seen with catalytically inactive mutants of these enzymes. There is evidence that EGF activation of MAP kinase depends on endocytosis of the EGF receptor (Vieira et al. 1996). In support of a role for PLD in internalization of the receptor, EGF-induced phosphorylation of MAP kinase was found to be inhibited by 1-butanol, but not 2-butanol (Shen et al. 2001). Further evidence of a role of PLD in endocytosis comes from studies of the recruitment of AP-2 proteins into the plasma membane and their targeting to endosomes (West et al 1997). These proteins are required for the formation of clathrin-coated vesicles which are involved in endocytosis. Addition of exogenous PLD to liver membranes or permeabilized NRK cells was found to promote the recruitment of AP-2 proteins to endosomes to the same extent as GTP(S, and the recruitment of AP-2 to endosomes and the plasma membrane was reduced by neomycin in the presence and absence of GTPγS.

PLD has also been reported to be involved in the assembly of AP-2-containing clathrin coats on lysosomes (Arneson et al. 1999). This is thought to be due to the effects of PA on PIP_2 synthesis. Fusion of early endosomes may also involve PLD since this process is blocked by 1-butanol, but not 2-butanol, and is stimulated by exogenous PLD (Jones & Clague 1997). In contrast to AP-2 proteins, the recruitment of COP proteins to endosomal membranes does not seem to involve PLD (Gu & Gruenberg, 2000). For example, it was observed that addition of PLD or PA to endosomes did not promote COP binding, whereas GTP(S had a strong effect through activation of ARF.

An interesting proposed new role for PLD is in trafficking of the GLUT4 glucose transporter (Emoto et al. 2000). This translocation is involved in insulin-stimulated glucose transport in muscle and adipose tissue. Using Myc-tagged PLD1, the expressed enzyme was shown by immunofluorescence and confocal microscopy to partly colocalize with GLUT4 in 3T3-L1 adipocytes and CHO cells. In addition, microinjection of PLD into adipocytes markedly potentiated the effect of insulin on GLUT4 translocation to plasma membranes (Emoto et al. 2000). These results were supported by the observation that 1-butanol, but not 2-butanol, inhibited insulin-stimulated 2-deoxyglucose transport in 3T3-L1 adipocytes. The precise role of PLD in stimulating GLUT4 trafficking remains to be defined.

7.3
Role in Mitogenesis

Early studies implicated PLD and PA in the regulation of mitogenesis. However, many of these experiments involved addition of PA to cells and were complicated by the presence of LPA in the PA preparations. Some early experiments also involved incubation of intact cells with plant or bacterial PLD, but this is also problematic since it can produce LPA as a result of the action of these enzymes on LPC (van Dijk et al. 1998). Therefore these types of experiments do not provide conclusive evidence for a role of PLD or PA in cellular processes. Other approaches have examined the correlation between cellular PA levels and DNA synthesis or have manipulated the PA level through the use of primary alcohols or inhibitors of PAP e.g. propranolol. In Swiss 3T3 cells, mitogenic concentrations of sphingosine induced early increases in PA, whereas structurally related analogs did not affect either process (Zhang et al. 1990). Stimulation of Balb/c3T3 cells with PDGF caused a rapid, large and prolonged increase in PA, whereas the accumulation of DAG was biphasic (Fukami & Takenawa 1992). Propranolol enhanced the increase in PA, whereas an inhibitor of DAG kinase had no effect, indicating that the PA was produced mainly through PLD action. Treatment of the cells with exogenous PLD or PA, from which LPA had been removed, stimulated DNA synthesis (Fukami & Takenawa 1992), whereas PLC was less effective. These findings suggest that PA is more effective than DAG in regulating mitogenesis in these cells. Kondo et al. (1992) also found evidence for a role of PLD in the control of DNA synthesis in vesicular smooth muscle cells. Exogenous PLD from *S. chromofuscus* stimulated mitogenesis in these cells, and the effect was enhanced by insulin. Downregulation of PKC did not affect the action of PLD, suggesting that it was not due to the generation of LPA.

The proliferative action of thrombin on MC3T3-E1 osteoblasts has been proposed to involve PLD (Suzuki et al. 1996a). This is based on the observations that thrombin causes choline release from these cells and that propranolol inhibits DAG formation and DNA synthesis. However, more studies are required to support this proposal. A role for PLD in the mitogenic action of arginine vasopressin (AVP) on glomerular mesangial cells is also indicated. AVP increased mitogenesis and activated PLD in these cells (Kusaka et al. 1996). Preincubation of the cells with the putative PLD inhibitors 2,3-diphosphoglycerate and carbobenzyloxy-leucine-tyrosine-chloromethylketone reduced the activation of both PLD and MAP kinase by AVP. Exogenous bacterial PLD also stimulated DNA synthesis. The stimulatory effects of AVP and bacterial PLD on thymidine incorporation into DNA were also blocked by 2,3-diphosphoglycerate. All these results support a role for PLD in the mitogenic action of AVP. However, the specificity of the PLD inhibitors may be questioned, and also the mode of action of exogenous PLD.

Another exploration of the role of PLD in mitogenesis induced by angiotensin II in vascular smooth muscle cells utilized 1-and 3-butanols to diminish PA accumulation (Wilkie et al. 1996). 1-butanol caused a significant decrease in thymidine incorporation into DNA, whereas 3-butanol was ineffective. In another exploration of the role of PLD in the mitogenic action of angiotensin II, the agonist and exogenous PLD and PA were all shown to increase DNA synthesis in vascular smooth muscle cells (Freeman 2000). However, propranolol blocked the stimulation by all these agents even though it increased PA, and an inhibitor of DAG lipase attenuated the incorporation of thymidine into DNA. Thus, in this system, DAG derived from PA may be an important signal for mitogenesis, as suggested also by the studies of Suzuki et al. (1996). In contrast, a recent report indicates the importance of PA in mitogenesis. This utilized a different approach to defining a role of PLD in mitogenic signaling, involving the use of NIH-3T3 cells stably overexpressing PLD1 (Hong et al. 2001). These cells showed enhanced PLD and MAP kinase activity in response to LPA (Fig. 24). The activation of MAP kinase was decreased by 1-butanol and enhanced by propranolol, indicating the involvement of PA, but not DAG. Evidence of the involvement of PLD

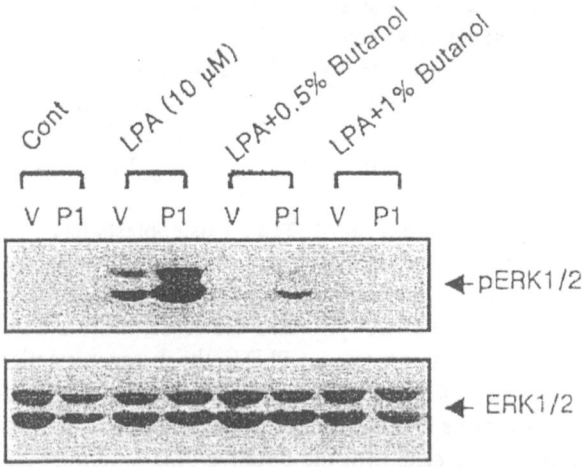

Fig. 24. Enhanced activation of MAP kinase in fibroblasts overexpressing PLD1. Vector-expressing (V) or PLD1-overexpressing (P1) NIH 3T3 cells were incubated with LPA in the presence of absence of two concentrations of 1-butanol. The activation of MAP kinase (ERK1) was assessed with phospho-specific anti-ERK antibodies. From Hong et al. (2001) by permission of the authors and publisher

and PA in the stimulation of MAP kinase by interleukin-11 in 3T3-L1 cells has also been presented (Siddiqui & Yang 1995).

A role for PA in the EGF induced transition of A431 epithelial cells from the G_2 phase of the cell cycle to mitosis has been proposed (Kaszkin et al. 1992). The cells responded to EGF with formation of PA which dose-dependently correlated with a delay in transition from G_2 phase. Treatment with PMA, which increased the PA level, and with exogenous PLD also decreased the G_2-M transition rate.

7.4
Role in Superoxide Production

Neutrophils play a major role in host defense against microorganisms. One of the responses is the generation of toxic O_2 metabolites, including the superoxide anion (O^-_2), and PLD has been implicated in this process. For example, agonists that elicit this response, e.g. chemoattractants, phagocytic particles and cytokines, also stimulate PLD activity (Billah et al. 1989; Agwu et al. 1989; Pai et al. 1988). In addition, early experiments demonstrated that PA elicited NADPH-dependent production of superoxide (O^-_2) in membrane extracts of neutrophils (Bellavite et al. 1988) or when this lipid was added to intact neutrophils (Ohtsuka et al. 1989). In neutrophils primed with tumor necrosis factor α, FMLP caused accumulation of PA and this was highly correlated with O^-_2 production (Bauldry et al. 1991). Through the use of neutrophils or HL60 cells labeled with [3H]hexadecyl-2-lyso-sn-glycero-3-phosphocholine ([3H]alkyl-lysoPC) and alkyl-[32P]lysoPC, the PA was shown to be formed by PLD activation (Agwu et al. 1989; Pai et al. 1988; Billah et al. 1989; Bauldry et al. 1991), and ethanol inhibition of PA formation led to a corresponding reduction of O^-_2 generation (Bonser et al. 1989; Bauldry et al. 1991). Similar findings were made by Agwu et al. (1991) who also measured NADPH oxidase activity in cell-free extracts. Propranolol at low concentrations enhanced both PA accumulation and oxidase activity, but at higher concentrations the stimulation was lost (Agwu et al. 1991). Another group reported that PA produced by PLD, and not DAG, was functionally linked to the oxidase (Rossi et al. 1990). Didecanoyl PA activated NADPH oxidase in vitro, in agreement with earlier findings. In a more extensive study utilizing permeabilized neutrophils, further evidence was obtained for a role of PA (Bauldry et al. 1992).

Other studies demonstrated that DAG could also play a role in the activation of NADPH oxidase. For example, Burnham et al. (1990) observed that DAG could synergize with SDS to activate the enzyme, and Qualliotine-

J. H. Exton

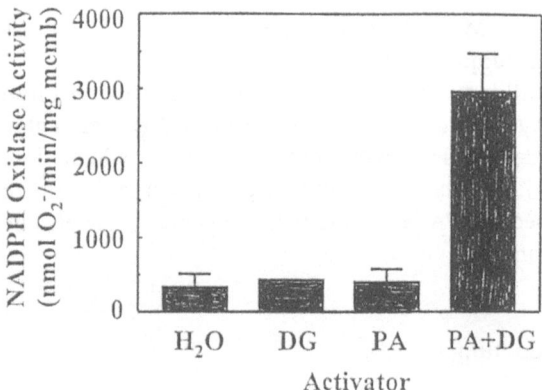

Fig. 25. PA and DAG synergize to activate NADPH oxidase in vitro. Cytosolic and membrane fractions from neutrophils were incubated with a reaction mixture containing the lipids shown. NADPH oxidase was measured as described by Qualliotine-Mann et al. (1993)

Mann et al. (1993) reported a synergism between PA and DAG (Fig. 25). This was confirmed in a later study (Erickson et al. 1999) which demonstrated that the activation of the oxidase by PA was not due to its conversion to DAG. Evidence was then presented that PA acted through a PA-regulated protein kinase (McPhail et al. 1995), which phosphorylated the p47[phox] component of NADPH oxidase (Waite et al. 1997). In later work, the p22[phox] subunit was also shown to be phosphorylated by the PA-activated kinase and also by conventional PKC isozymes (Regier et al. 1999). Phosphorylation of the p22[phox] subunit was observed in neutrophils stimulated by FMLP and other agonists. Pretreatment of the cells with ethanol produced large decreases in both NADPH oxidase activity and p22[phox] phosphorylation (Regier et al. 2000). However, these responses were not altered by ethanol in cells stimulated by PMA, indicating the existence of both PLD-mediated and PKC-mediated pathways for p22[phox] phosphorylation and NADPH oxidase activation.

Despite the preceding evidence that NADPH oxidase can be controlled by protein kinases sensitive to PA and DAG, there is also evidence that these lipids can also act by a non-phosphorylation mechanism (McPhail et al. 1995). Thus, in a cell-free system of purified recombinant NADPH oxidase components, activation of the complex by PA plus DAG was ATP-independent and not affected by staurosporine (Palicz et al. 2001). In summary, stimulation of neutrophils by FMLP and other agonists causes PA accumulation as a result of PLD activation. This PA in combination with

DAG can activate NADPH oxidase in neutrophil by phosphorylation-dependent and -independent mechanisms.

7.5
Role in Actin Cytoskeleton Rearrangements

Alterations in the actin cytoskeleton are important in cell shape changes and motility, chemotaxis, cell division, endocytosis and secretion. The changes in cell shape, motility, migration and metastasis involve activation of members of the Rho family of GTPases. For example,Rho mediates stress fiber formation and focal adhesion formation in fibroblasts and causes neurite retraction and rounding up in neurons, whereas Rac is involved in membrane ruffling and the generation of lamellipodia, and activation of Cdc42 induces filopodia formation (Nobes and Hall, 1994).

The involvement of PLD in Rho-mediated changes in the actin cytoskeleton was first suggested in studies of the effects of exogenous bacterial PLD and PA on actin polymerization and reorganization in IIC9 fibroblasts (Ha and Exton 1993a). These agents induced stress fiber formation similar to that caused by α-thrombin and also increased the F-actin content of the cells. In contrast, dioctanoylglycerol and PC-PLC from *B. cereus* were without effect. However, the possible role of LPA contamination or generation was not ruled out. In further work, LPA was found to cause a very rapid activation of PLD in IIC9 cells which was accompanied by an equally rapid increase in F-actin (Ha et al. 1994). However, the polymerization of actin was observed with lower concentrations of LPA than was PLD activation, suggesting that either a small increase in PA was required for the actin effect or the two processes were dissociated. Support for a role of PLD in actin stress fiber formation came from a study of cytoskeleton changes in endothelial cells (Cross et al. 1996). In agreement with Ha and Exton (1994) addition of PA induced stress fiber formation in these cells, whereas dioctanoylglycerol caused only a small effect. As expected, LPA also induced this cytoskeletal change and the effects of both PA and LPA were blocked by C3 exoenzyme, indicating the involvement of Rho. A role for PLD in regulation of the actin cytoskeleton was suggested by the observations that LPA activated PLD and that 1-butanol inhibited LPA-induced formation of both PA and actin stress fibers, whereas 2-butanol was without effect (Cross et al. 1996). In these experiments, PA was shown by thin layer chromatography to be free of LPA and not to activate PLD. Furthermore, its action on stress fibers was not inhibited by 1-butanol.

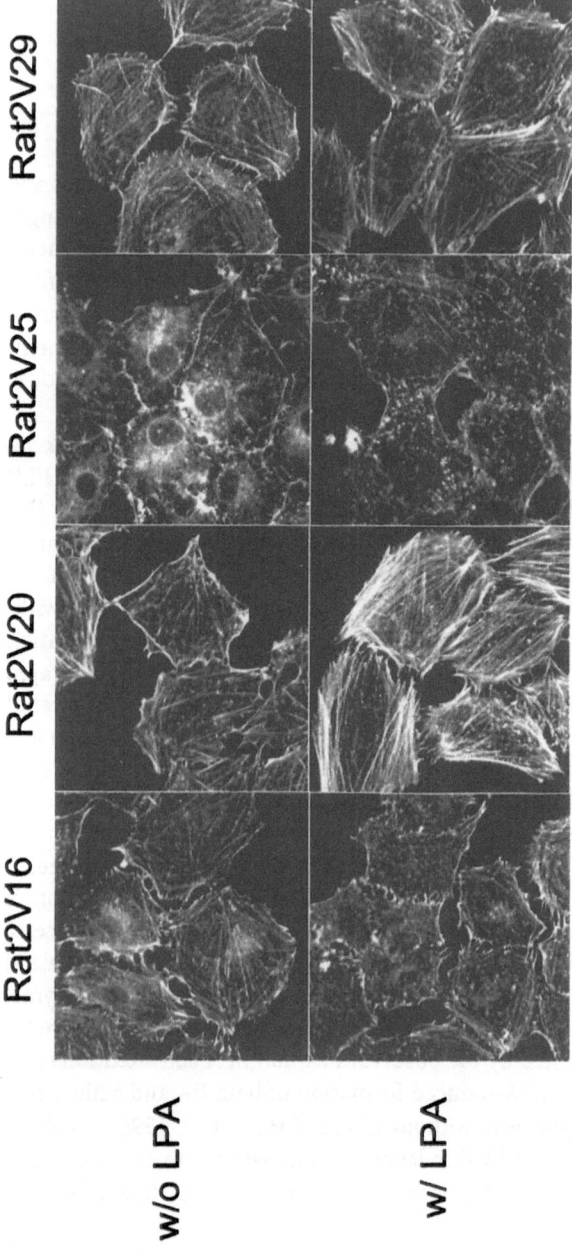

Fig. 26. Fibroblasts with reduced PLD activity show loss of stress fiber formation in response to LPA. Rat-2 fibroblast clones stably expressing inactive V5 epitope-tagged rPLD1 and showing reduced PLD activity were treated with LPA, and stress fiber formation assessed by staining with Texas red X-phalloidin and confocal microscopy. The V16, V25 and V29 clones showed expression of V5-rPLD1 and reduced PLD activity. The V20 clone did not express V5-rPLD1 and showed normal PLD activity. From Kam & Exton (2001) by permission of the authors and publisher

A recent study (Kam and Exton, 2001) has provided more evidence for a role of PLD in actin stress fiber formation in cells. These workers generated Rat1 fibroblast lines stably expressing catalytically inactive V5 epitope-tagged PLD1. These cell lines exhibited reduced basal and PMA- or LPA-stimulated PLD activity and failed to show normal actin stress-fiber and α-actinin translocation responses to LPA (Fig. 26). In contrast, the cells showed normal membrane ruffling in response to PMA and focal adhesion formation was intact, as revealed by immunostaining of vinculin. Since the activation of RhoA and of Rho kinase by LPA was unimpaired in the PLD-deficient cells, it was concluded that PLD was involved in downstream events from Rho kinase (Kam and Exton, 2001). Because PIP_2 affects the function of many actin-associated cytoskeletal proteins (Sechi & Wehland 2000), the authors proposed that the role of PLD was to generate PA which then acted on type 1 PI 4-P 5-kinase (Jenkins et al. 1994; Ren et al. 1996; Moritz et al. 1992) to produce a local increase in PIP_2. Further work is needed to support this hypothesis.

An association of PLD with the actin cytoskeleton was indicated by Iyer and Kusner (1999). These workers found that PLD activity was associated with the detergent-insoluble fraction of promonocytic leukocytes. This association was greatly increased by pre-incubation with GTPγS. The fraction was enriched in the cytoskeletal proteins F-actin, talin, paxillin and α-actinin, and PLD1, RhoA and ARF were also detected. A more direct interaction of PLD with actin was shown by Lee et al. (2001). Using mass spectrometry, these workers identified a major PLD2-binding protein in rat brain as β-actin. This protein was also found to bind directly to PLD2 and PLD1 and to inhibit their activity. Immunocytochemistry and co-immunoprecipitation studies showed that both PLD isozymes interacted with β-actin and with the actin cytoskeleton in intact cells. In summary, all these results support a close association between PLD and the actin cytoskeleton.

7.6
Role in Lysophosphatidic Acid Formation

LPA is an intermediate in the biosynthesis of triacylglycerol and certain phospholipids. It is also an important extracellular messenger which interacts with certain EDG receptors which are present on the surface of many cells and are coupled to heterotrimeric G proteins (Contos et al. 2000). Thus LPA produces proliferative and morphological effects that are due to changes in the activity of adenylyl cyclase, PLC, PLA_2, MAP kinase and Rho (Kranenburg & Moolenaar 2001). LPA is released from activated platelets and probably other cells, and is a major mitogen in serum (Eichholtz et al. 1993).

There are several ways by which LPA could be released from cells. One is by action of phospholipase A_2 on PA to yield 1-acyl LPA, which is the major form of LPA (Fig. 27). 2-acyl LPA can also be produced through the action of PLA_1. This form is equipotent with 1-acyl LPA on EDG2 and EDG4 receptors and more potent on EDG7 receptors (Bandoh et al. 2000). An alternative pathway for LPA production is through action of lysophospholipase D on 1- or 2-acyl lyso PC or other lysophospholipids to yield 1-acyl LPA or 2-acyl LPA. PLA_2 is present in numerous isoforms in mammalian cells (Dennis 1994) and PLA_1 has a widespread distribution (Pete et al. 1994). Secreted forms of PLA_2 could act on PA and other phospholipids in the outer leaflet of the plasma membranes to cause the release of LPA or formation of lysophospholipids, but it is unclear that PLA_1 could act in this manner. The nature of the lysophospholipase D that acts on lysoPC (vanDijk et al. 1998) is unclear. Bacterial forms of PLD exhibit lysophospholipase D activity, but this has not been demonstrated for mammalian PLD isoforms. If the principal route of LPA formation is from PA via phospholipase A action, then PLD is probably involved through its action on PC to yield PA.

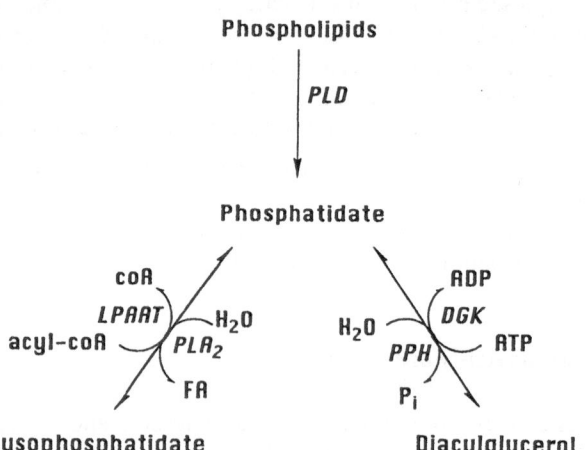

Fig. 27. Metabolism of PA to LPA and DAG. The figure shows conversion of PA to LPA by PLA2 and the reverse reaction catalyzed by lysophosphatidate acyltransferase (LPAAT), and the conversion of PA to DAG by phosphatidate phosphohydrolase (PPH) and the reverse reaction catalyzed by diacylglycerol kinase (DGK)

7.7
Other Cellular Roles

Many other functions have been ascribed to PLD. One is in cellular senescence (Venable et al. 1994). In this study, senescent human diploid fibroblasts were noted to have impaired DAG accumulation in response to serum. The defect was traced to an impairment in PLD activation, and it was suggested that this was due to the increase in ceramide observed in these cells.

Another role for PLD is in the turnover of PC (Tronchère et al. 1995). In neutrophils stimulated with FMLP, there was breakdown of PC to PA mediated by PLD, but no stimulation of PC synthesis. However if cytochalasin B was added with FMLP, there was conversion of PA to DAG and incorporation of choline into PC and translocation of the rate-limiting enzyme for PC synthesis (CTP: phosphocholine cytidyltransferase) from the cytosol to membranes. Further work indicated that only DAG derived from PC via PA could promote cytidyltransferase translocation and subsequent PC synthesis (Tronchère et al. 1995).

Roles for PLD-generated PA in other cellular responses have been proposed. For example, priming of neutrophils with platelet-activating factor caused PLD activation and the binding of serum-opsonized particles, whereas ethanol greatly inhibited this binding and also PA formation and the respiratory burst (Tool et al. 1999). The PAP inhibitor propranolol, enhanced the binding of the particles and addition of cell-permeant PA resulted in CD11b/CD18 integrin-dependent adhesion to the particles and fibronectin. The authors concluded that PLD-derived PA was involved in altering the affinity of the integrin for its ligands.

The role of PLD in the supply of choline for acetylcholine synthesis in the brain remains speculative. Free choline is present at low concentrations in the brain, and the turnover of acetylcholine suggests that the reuptake of choline is insufficient for continued synthesis of this neurotransmitter (Klein et al. 1995). Generation of the extra choline by PLD activity remains an attractive possibility and the agonist that stimulates PLD could be acetylcholine itself. However, the site of PLD activation and choline release, e.g. presynaptic or glial cells, and the agonists involved remain unclear.

A role for PLD in AVP-induced Ca^{2+} spiking in vascular smooth muscle cells has been proposed (Li et al. 2001). Treatment of the cells with a putative PLA_2 inhibitor (ONO-RS-082) led to inhibition of PA formation, but not that of arachidonic acid. Furthermore, exogenous PLD induced Ca^{2+} spiking, whereas arachidonic acid was without effect. 1-butanol, but not 2-butanol, suppressed the spiking caused by AVP, but had no effect on $BaCl_2$-stimulated spiking. Based on inhibitor data PKC was proposed to be downstream from PLD in the signaling pathway involved in AVP-induced Ca^{2+} spiking.

7.8
Other Targets of Phosphatidic Acid

Cellular targets for PA have been described in some preceding sections. These include NADPH oxidase, Golgi and other membranes involved in vesicle trafficking, PI 4-P 5-kinase and protein kinases. In this section, additional PA targets will be described, but their in vivo significance remains unclear. Early work identified RasGAP as a target (Tsai et al. 1989) and the activation of Ras due to inhibition of its GAP activity by PA was thought to underlie the mitogenic action of this lipid. Inhibition of Rac interaction with Rho GDI has also been reported (Chuang et al. 1993) and n-chimaerin an activator of RacGAP, is stimulated by PA (Ahmed et al. 1993). However, there have been no reports of PLD action on Rac activity in intact cells. PA binds strongly and selectively to Raf 1 kinase, and inhibition of PA formation by treatment of MDCK cells with ethanol reduces membrane translocation of Raf 1 (Ghosh et al. 1996). PA-mediated recruitment of Raf 1 was reported by another group (Rizzo et al. 2000) who showed that the interaction was independent of Ras.

In HL60 cells, production of PA via activation of endogenous PLD by PMA or treatment with exogenous PA or PLD from *S. chromofuscus* results in tyrosine phosphorylation of 100–115 kDa proteins (Ohguchi et al. 1997). The phosphorylation induced by PMA is reduced by ethanol or butanol, suggesting the involvement of PLD. PA-dependent tyrosine phosphorylation of cellular proteins was also observed in studies with neutrophils (Sergeant et al.2001). Similar to what was found in HL60 cells, PMA-induced phosphorylation was decreased by primary alcohols. The tyrosine kinase(s) responsible for these effects has not been identified.

In contrast to the preceding results, PA has been found to stimulate a specific protein tyrosine phosphatase termed PTP1C (also known as SH-PTP1, HCP or SHP-1). This contains two SH2 domains and is very strongly stimulated by anionic phospholipids including PA, which is the most effective activator (Zhao et al. 1993). A later study showed that PA bound to a site at the C-terminus of the enzyme (Frank et al. 1999). PTP1C was found to be associated with the EGF receptor in A431 cells (Tomic et al. 1995), but showed little activity towards the autophosphorylated receptor unless PA was added. The lipid also enhanced the association of PTP1C with the receptor and stimulated its dephosphorylation in intact cells. The significance of these findings for EGF signaling remains to be established.

PA can interact with certain PLC isozymes in vitro. PA was first reported to activate PLC activity associated with platelet membranes (Jackowski & Rock 1989). In a more recent study, it was found to alter the kinetics of PLC-γ1 by decreasing the cooperativity index with little change in the association with substrate micelles (Jones et al. 1993). However, PA increased the activity of tyrosine phosphorylated form of the enzyme by decreasing the Km 10-fold. PLC-β1 is also stimulated by PA, which increases its association with unilamellar vesicles (Litosch 2000). PA increases the enzymatic activity in the presence of PIP_2 and Ca^{2+} and decreases the Ca^{2+} concentration required for stimulation.

Certain isoforms of cAMP-specific phosphodiesterase are regulated by PA. Thus PDE4A5 is stimulated by PA in vitro (Fig. 28), and treatment of thymocytes with DAG kinase inhibitors to decrease the PA level caused a reduction in concanavalin A-stimulated cAMP-specific PDE activity (El Bawab et al. 1997). PA extracted from ConA-stimulated cells was more effective in stimulating PDE4A5 than that extracted from unstimulated cells, and PA containing unsaturated fatty acids was more effective than that containing saturated fatty acids. PA has also been reported to bind to and activate PDE4D3 (Grange et al. 2000). Treatment of MA10 Leydig tumor cells overexpressing PDE4D3 with propranolol caused accumulation of PA. This was accompanied by an increase in PDE activity and a decrease in cAMP and protein kinase A activity. Furthermore in FRTL5 thyroid cells, which natively express PDE4D3, agents that induce PA accumulation decreased cAMP and CREB phosphorylation. These results suggest a role for PA in regulating cAMP signaling in cells.

The existence of a PA-stimulated Ser/Thr protein kinase in neutrophils was alluded to in Section 7.4. This activity was also detected in a variety of tissues and shown to be independent of Ca^{2+} and separate from PKC (Bocckino et al. 1991). Another lipid-dependent protein kinase was characterized in platelets (Khan et al. 1994). This was selective for PA and was shown to be Ca^{2+}-independent and not a PKC isozyme. Certain PKC isozymes can be activated by PA. Thus PA can substitute for PS as an activator of the enzyme (Hannun et al. 1986;

Epand and Stafford 1990). In a detailed examination of the effects of PA on different PKC isozymes, Limatola et al. (1994) observed, in the absence of Ca^{2+}, that PKCζ was most strongly activated by PA, although PKCα was also activated by PA in the presence of Ca^{2+}. PA caused a shift in the electrophoretic mobility of PKCζ due to its binding to the enzyme. Concerning the effects of PA on protein Ser/Thr phosphatases, this lipid has been reported to be a potent and selective inhibitor of PP1, but not of PP2A or PP2B (Kishikawa et al. 1999). PA acts on the catalytic subunit of PP1(and is able to

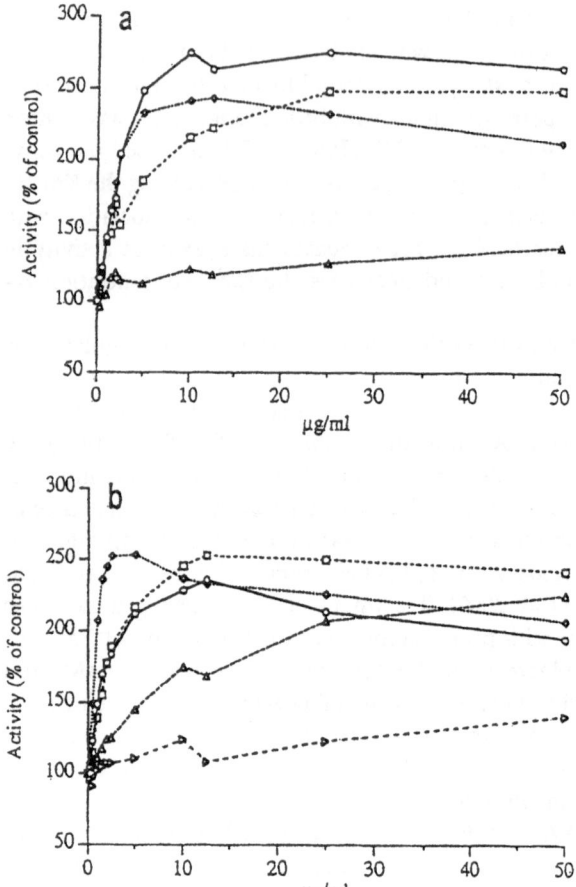

Fig. 28a, b. Activation of cAMP phosphodiesterase PDE4A5 by PA. Recombinant PDE4A5 prepared in Sf9 cells was incubated with different molecular species of PA and the phosphodiesterase activity measured by the conversion of [^{3}H]cAMP to [^{3}H]adenosine in the presence of 5'nucleotidase. Most PA species were very effective, except for distearoyl PA (Δ in panel a) and 1-palmitoyl-2-oleoyl PA and dipalmitoyl PA (Δ and in panel b). From El Bawab et al. (1997) by permission of the authors and publisher

counteract the stimulatory effects of ceramide. The possibility that PA may regulate protein phosphorylation/dephosphorylation in vivo needs further exploration.

A recent report has proposed a unique role for PA in the regulation of translation (Fang et al. 2001). This has identified the rapamycin-sensitive protein mTOR/FRAP as a target of PA. mTOR regulates translation initiation through effects on the initiation factor binding protein 4E-BP1 and the protein kinase S6K1 which phosphorylates ribosomal protein S6. PA was found to stimulate S6K1 activity and to induce 4E-BP1 hyperphosphorylation (Fang et al. 2001). Furthermore, PA interacted directly with the rapamycin binding domain of mTOR. It was proposed that PA mediates activation of mTOR by mitogenic stimuli and that rapamycin inhibits mTOR signaling by blocking PA binding. It will be interesting to see how this scheme relates to other mechanisms of mTOR regulation.

7.9
Diacylglycerol Production from Phosphatidylcholine and its Role in Protein Kinase C Regulation

An important issue in understanding PLD effects in vivo is whether PA and DAG derived from the action of PLD on PC are functionally different from those produced by PLC action on PIP_2. Banschbach et al. (1981) first noted that the DAG accumulating in mouse pancreas incubated with acetylcholine had a different fatty acid composition from that in PI, indicating that it came from another source. Although these workers suggested that this was triacylglycerol, the composition of the DAG resembled that of PC in the pancreas. Cockcroft & Allan (1984) also noted that the fatty acid composition of PA and DAG in neutrophils treated with FMLP was different from that of PI, and also suggested that these lipids were derived from another source. Bocckino et al. (1985) later demonstrated that two chromatographic peaks of DAG were produced by hepatocytes stimulated with AVP, and that these differed in their fatty acid composition. However, neither peak resembled PI in composition, and it was suggested that PC might be a source. This was confirmed in later studies of PA formation and PLD activation in hepatocytes stimulated with several agonists (Bocckino et al. 1987a; 1987b). Chemical measurements showed a rapid accumulation of PA that preceded a rise in DAG in the stimulated cells, indicating that the PA was unlikely to have arisen from DAG kinase activity. Furthermore, the fatty acid composition of the PA and DAG was noted to resemble that of PC and not PI. Studies utilizing the transphosphatidylation reaction to measure PLD activity then demonstrated that the agonists activated PLD in the cells, indicating that PLD activity was the probable source of the PA. Measurements of choline release from GTPγS- and hormone-stimulated liver plasma membranes indicated that PC was the substrate of PLD (Irving & Exton1987; Bocckino et al. 1987a). This was confirmed by isotope studies and analyses of the molecular species of DAG and phospholipids in hepatocytes stimulated with AVP (Augert et al. 1989). The major species of DAG in the stimulated hepatocytes were found to be 16:0/18:2, 18:0/20:4, 16:0/18:1, 16:0/20:4 and 18:0/18:2 (Augert et al. 1989). This pattern resembled that of PC, but not PI, extracted from the hepatocytes. Wright et al. (1988) analyzed the kinetics of DAG production in IIC9 fibroblasts stimulated with α-thrombin and EGF. Treatment with α-thrombin induced an early phase of DAG formation peaking at 15s and a late phase peaking at 5 min. With low concentations of this agonist or with EGF, only the slow phase was observed. Since the rapid phase corresponded with IP$_3$ formation, it was proposed to be derived from PIP$_2$ hydrolysis, whereas the second phase was derived from another source. Very similar results were obtained later by Ha & Exton (1993b) who analyzed the effects of α-thrombin and PDGF on DAG levels in IIC9 fibroblasts (Fig. 29). In detailed analyses of the molecular species of DAG and phospholipids in IIC9 fibroblasts, Pessin & Raben, (1989) and Pessin et al. (1990) concluded that phosphoinositides contributed significantly to the early phase of DAG generated by α-thrombin stimulation, whereas PC contributed most of the DAG measured at 5 min and 1 h. In the case of EGF stimulation, PC was the primary source, as also indicated by the sustained release of choline. Lee et al. (1991) studying muscarinic activation of SK-N-SH neuroblastoma cells noted that the molecular species of DAG accumulating at 5s resembled those in phosphoinositides, whereas those measured at 10 min or longer resembled those in PC.

In another study of the sources of DAG in agonist-stimulated cells, Huang & Cabot (1990) incubated REF52 fibroblasts with [^{3}H]myristic acid, which labels predominantly PC, and [^{3}H]arachidonic acid, which is incorporated into most phospholipids. In cells labeled with arachidonic acid and stimulated with AVP, there was a rapidly generated peak of DAG and a slower increase in PA, consistent with initial formation of DAG through PIP$_2$ hydrolysis. On the other hand, in myristate-labeled cells, PA increased rapidly, but DAG did not change until later, in accord with PA formation from PC by PLD. Measurements of inositol phospholipid changes and of phosphatidylethanol formation from specifically labeled PC confirmed these conclusions (Huang & Cabot 1990).

Support for the view that agonists generate DAG in cells by both PLC-catalyzed breakdown of PIP$_2$ and PLD-mediated hydrolysis of PC has come from other studies utilizing isotopic labeling and molecular species analysis in a variety of cell types stimulated with different agonists (Martinson et al.1989; Price et al. 1989; Matozaki & Williams 1989; Kennerly 1990; MacNulty et al. 1990; Lee et al. 1991; Plevin et al. 1991; Divecha et al. 1991; van Blitterswijk et al. 1991a; 1991b; Sebeldt et al. 1992; Fällman et al. 1992; Hermans et al. 1996). In general, these studies showed that DAG was derived mainly from PIP$_2$ in the early phase of agonist stimulation, and from PC in

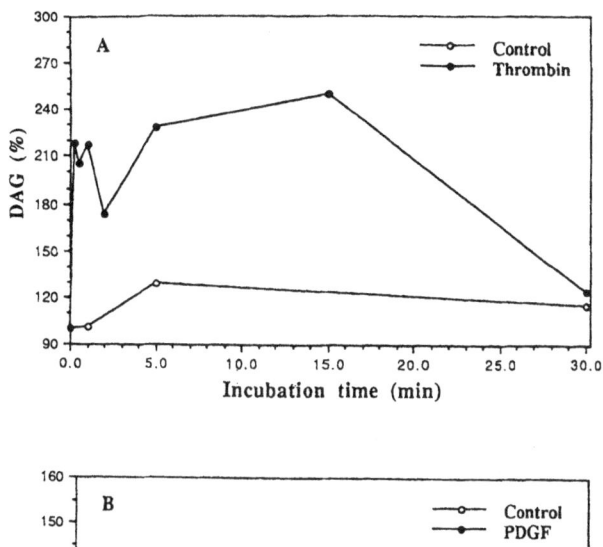

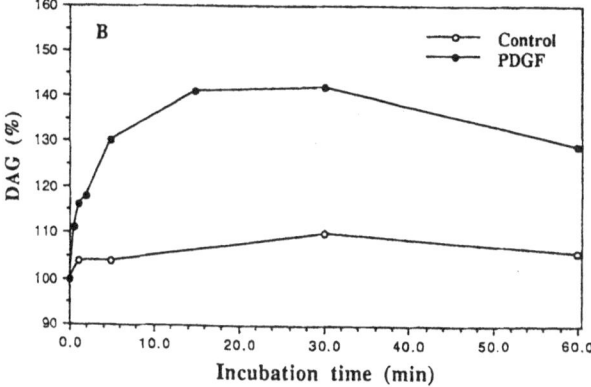

Fig. 29A, B. Time courses of formation of DAG in fibroblasts stimulated with α-thrombin (A) or PDGF (B). IIC9 fibroblasts were labeled with [^{3}H]myristic acid and incubated with α-thrombin or PDGF. [^{3}H]DAG was measured at the indicated times by thin layer chromatography. From Ha & Exton (1993b) by permission of the authors and publisher

the late phase. Likewise PA was initially produced mainly by DAG kinase action on PIP$_2$ and later by PLD action on PC.

Pettit et al. (1997) also looked at the composition of DAG and PA in endothelial cells stimulated with LPA. Although these workers found that LPA produced no significant changes in individual DAG species, the pattern of molecular species was similar to that observed by Augert et al. (1989), Pessin & Raben (1989), Pessin et al. (1990), Lee et al. (1991) or Hermans et al. (1996) in cells stimulated for a short time with agonists. It reflected formation predominantly from phosphoinositides. The fatty acid composition of PA and phosphatidylbutanol in Swiss 3T3 cells stimulated with bombesin in the presence of butanol and in endothelial cells treated with LPA and butanol was also examined by Pettit et al. (1997). These workers found a predominance of 16:0, 18:0 and 18:1 fatty acids with almost undectable levels of 16:1, 18:2, 20:4 and 22:6 fatty acids, and concluded that PA and phosphatidylbutanol derived from PLD action had a very different fatty acid composition from DAG generated by PLC. Although this conclusion is correct, their data make it difficult to identify the phospholipid acted on by PLD since all mammalian phospholipids contain significant amounts of arachidonic acid and other polyunsaturated fatty acids in their sn-2position (Kuksis et al. 1968; Holub & Kuksis 1978; Smith & Jungalwala 1981; Mahadevappa & Holub 1982; Patton et al. 1982; Takamura et al. 1987; MacDonald & Sprecher 1989; Augert et al. 1989; Pessin & Raben 1989; Pessin et al. 1990; Lee et al. 1991). This presence of arachidonic acid in this position in PC is the reason why activation of PLA2 leads predominantly to the release of arachidonic acid.

The cellular role of DAG accumulating in the late phase of agonist stimulation of cells remains unclear (Wakelam 1998). It was originally postulated to produce prolonged activation of PKC, as opposed to transient activation due to PIP_2 hydrolysis (Bocckino et al. 1985; Price et al. 1989; Matozaki & Williams 1989; Exton 1990; Fällman et al. 1992). It was also proposed that the prolonged increase in DAG was related to the mitogenic action of EGF (Wright et al. 1988). The view that PC-derived DAG could activate PKC was supported by several lines of evidence including analyses of its molecular species, which were found to be effective in activating PKC in vitro (Go et al. 1987). In MDCK cells, activation of α_1-adrenergic receptors resulted in hydrolysis of both PIP_2 and PC (Slivka et al. 1988). Treatment with neomycin to suppress PIP_2 breakdown did not significantly alter DAG production, indicating that it mainly came from PC. More importantly, neomycin did not inhibit the membrane translocation of PKC, implying that DAG derived from PC was capable of activating PKC. Furthermore, Fällman et al. (1992) showed that the sustained increase in DAG due to complement receptor-mediated activation of phagocytosis in neutrophils was associated with phosphorylation of MARCKS a well-known substrate of PKC. On the other hand, several groups have reported that the second phase of hormone-stimulated DAG accumulation does not activate PKC (Martin et al. 1990; Leach et al. 1991; Baldassare et al. 1992). However, these investigators studied mainly conventional (Ca^{2+}-dependent) PKC isozymes. In a detailed study of the activation of PKC isozymes in IIC9 fibroblasts stimulated with two different agonists Ha & Exton (1993b) showed that α-thrombin caused a rapid, but transient, translocation of PKCα to the membrane fraction

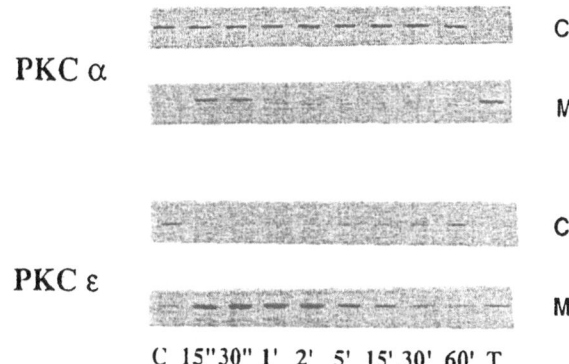

$$PKC \alpha$$

$$PKC \varepsilon$$

C 15"30" 1' 2' 5' 15' 30' 60' T

Fig. 30. Translocation of PKC isozymes induced by α-thrombin in IIC9 fibroblasts. Cells were activated with α-thrombin for the indicated times or with PMA (T) for 15 min. Cytosolic (C) and membrane (M) fractions were isolated and Western blotted for PKCα and PKCε. From Ha and Exton (1993b) by permission of the authors and publisher

(Fig. 30). This coincided with transient increases in IP$_3$ and cytosolic Ca^{2+}. PKCε was also rapidly translocated, but its membrane association persisted for 1 hr (Fig. 30). In contrast, PDGF did not cause PKCα to translocate and induced a slow membrane association of PKCε. PKCα translocation was blocked by Ca^{2+} chelators and could be induced by 1,2-dioctanoylglycerol plus ionomycin, but by neither agent alone. In contrast, the DAG alone could cause translocation of PKCε (Ha & Exton 1993b). These findings indicate that PC hydrolysis alone can activate Ca^{2+}-independent PKC isozymes, whereas activation of Ca^{2+}-dependent isozymes is dependent on a concurrent elevation of both DAG and cytosolic Ca^{2+} i.e. PIP$_2$ hydrolysis.

Some cells contain significant amounts of ether-linked phospholipids and there have been some studies of the effects of agonists on the release of 1-alkyl-2-acylglycerol and 1-alkenyl-2-acylglycerol. In pancreatic acini stimulated with cholecystokinin, 1,2-diacylglycerol and 1-alkenyl-2-acylglycerol, but not 1-alkyl-2-acylglycerol, accumulated due to hydrolysis of phosphoinositides, PC and plasmenyl-PC (Hermans et al. 1996). Due to the lower content of plasmenyl-PC, the predominant species was 1,2-diacylglycerol derived mainly from PC, but both types of diradylglycerol contained large amounts of linoleic and arachidonic acids. Stimulation of neutrophils with FMLP caused the production of both 1-alkyl-2-acylglycerol and 1,2-diacylglycerol (Dougherty et al. 1989), but their kinetics differed. The diacylglycerol was generated more rapidly than the alkyl-acylglycerol. Although alkyl-acylglycerol showed a greater fold-increase, its basal and stimulated levels were less than those of diacylglycerol. In a study of diradylglycerol formation in mesangial cells, Musial et al. (1995) observed that endothelin-1 generated predominantly 1,2-diacyl species, whereas interleukin-1 produced ether-linked species (alkyl-acylglycerol and alkenyl-acylglycerol).

There is still some controversy about the effects of 1-alkyl-2-acylglycerols on PKC activation. As noted above, these diradylglyerols can be generated by agonist stimulation of certain cells. Several groups have reported that they are ineffective in activating PKC (Cabot & Jaken 1984; Ganong et al. 1986; Daniel et al. 1988; Musial et al. 2001). However, their reported effects on the stimulation of PKC by PMA or 1,2-diacylglycerol have been variable. Three groups reported that 1-alkyl-2-acylglycerols inhibited the stimulatory effect of 1,2-diacylglycerols (Bass et al. 1989; Daniel et al.1988; Musial et al. 1995). Furthermore, in mesangial cells, in which interleukin-1 generates 1-alkyl-2-acylglycerol and 1-alkenyl-2-acylglycerol, the cytokine caused inhibition of PKC activation by endothelin 1 (Musial et al. 1995). 1-Alkyl-2-acylglycerol also inhibited the priming of the respiratory burst induced by 1,2-diacylglycerol in polymorphonuclear leukocytes (Bass et al. 1989). In contrast to these findings, 1-alkyl-2-acylglycerol has recently been reported to enhance the activation of conventional PKC isozymes by PMA or 1-oleoyl-2-acetylglycerol in vitro (Slater et al. 2001). In summary, although ether-linked diradylglycerols appear to be unable to activate PKC, their effects on the activation of the enzyme by diacylglycerols are unclear.

8
Summary and Future Directions

Although much progress has been made in characterizing the enzymatic properties and regulation of mammalian PLDs, deficiencies in knowledge remain. In particular, information on the three-dimensional structure of mammalian PLDs is lacking, although the crystal structures of a plant and bacterial enzyme and of a low Mr member of the PLD superfamily have been defined. These have given information about the catalytic center and mechanism, but not how the enzyme is regulated. The catalytic center is formed by dimerization of the two HKD domains present in the enzyme, with the two histidines playing essential roles in catalysis as nucleophile and general acid in a two-step reaction. Mutagenesis studies have identified the PKC interaction site at the N-terminus and the Rho interaction site at the C-terminus, but it is not known where ARF interacts.

Both PLD1 and PLD2 are membrane-associated and appear to be localized to different regions in cells. PLD1 has a predominant perinuclear localization (?Golgi) but is located in other undefined intracellular structures and may be in the plasma membrane in some cells. PLD2 appears to be mainly in the plasma membrane. Both isozymes are subject to palmitoylation, which is not essential for activity, but is a factor in their membrane association. PLD1 is phosphorylated in unstimulated cells by an undefined Ser/Thr kinase. Phosphorylation is restricted to the N-terminal half of the enzyme and requires palmitoylation and association of the two HKD motifs. The phosphorylation has minimal effects on the intrinsic activity of the enzyme , but is involved in membrane association.

The conventional (Ca^{2+}-dependent) isozymes of PKC are major regulators of PLD1 in vivo and in vitro. PLD2 can be activated by PKC-dependent mechanisms in vivo, but is not affected by PKC isozymes in vitro. Surprisingly, the activation of PLD1 by PKC in vitro does not involve phospshorylation, but occurs by a protein-protein interaction and only involves the α- and β-isozymes of the kinase. The mechanism(s) by which PLD1 and PLD2 are activated by PKC in vivo have not been defined, but may or may not involve phosphorylation of the enzymes. PKC activation appears to be a major mechanism by which agonists acting on receptors coupled to G proteins or growth factors interacting with receptors encoding tyrosine kinase activity activate PLD in vivo.

Small G proteins of the Rho and ARF families activate PLD1 in vitro, but have little or no effect on PLD2. Of the Rho family, Rho is more effective than Rac and Cdc42. The G_{13} heterotrimeric G protein can activate Rho and hence PLD, and there is evidence that several agonists stimulate PLD by this mechanism. Although Rho can interact directly with PLD1 in vitro, other mechanisms of Rho activation of the enzyme may occur in vivo. PLD1 is activated by all mammalian ARFs in vitro, but the in vivo significance of ARF regulation is unclear. Class I ARFs are probably important in the regulation of the enzyme in Golgi, and ARF6 may be involved in membrane trafficking at the plasma membrane. Some reports have indicated a role for ARF in agonist regulation of PLD.

Tyrosine kinase activity plays a role in the regulation of PLD1 and PLD2, but it is unclear that this involves direct phosphorylation of the enzymes or of proteins involved in upstream events. Recently, Ral, a member of the Ras subfamily of small G proteins, has been implicated in the regulation of PLD. Ral may mediate some of the effects of Ras and growth factors on the enzyme. Ca^{2+} ions also play a role in the regulation of PLD, possibly through an action mediated by calmodulin. Ceramide, a product of sphingomyelin breakdown, is an inhibitor of PLD in vitro and in vivo. PI 3-kinase may play a role in the indirect regulation of the enzyme in some cell types.

An area needing much more work is the definition of the cellular functions of PLD. This is because many approaches have not been successful. The role of PLD in vesicle trafficking in Golgi remains controversial, although there is more support for its role in exocytosis. Evidence is also emerging for a role of PLD in endocytosis. It is still unclear what function PLD plays in the mitogenic response of cells to agonists. This is because the studies have mostly utilized inhibitors of uncertain specificity. There is more evidence that PA plays a role in superoxide production by neutrophils. Thus, PA has been reported to have in vitro and in vivo effects on the NADPH oxidase system. A role for PLD activity in the Rho-mediated rearragements of the actin cytoskeleton involved in stress fiber formation seems likely. This may involve changes in PIP_2, which is a regulator of several proteins involved in the response. It is likely, but not certain, that PA is the major source of LPA released from platelets and other cells. LPA has major effects throughout the body through its interaction with certain EDG receptors. Another product of PA metabolism is DAG. DAG derived from this source can activate novel (Ca^{2+}-independent) PKC isozymes but does not act on the conventional isozymes unless cytosolic Ca^{2+} is elevated. However, it is clear that PA itself can interact with many cellular proteins and alter their activities. These involve proteins that regulate small G proteins, protein kinases, protein phosphatases, phospholipases, phosphodiesterases, phospholipid kinases and other regulatory proteins. An important element of PLD research will be to relate these and other effects of PA to the cellular changes induced by PLD activation.

Acknowledgements. I would like to thank Judy Nixon and Anthony Couvillon for their great help in preparing this review article.

References

Abousalham A, Liossis C, O'Brien L, Brindley DN (1997) Cell-permeable ceramides prevent the activation of phospholipase D by ADP-ribosylation factor and RhoA. J Biol Chem 272:1069-1075

Agwu DE, McPhail LC, Chabot MC, Daniel LW, Wykle RL, McCall CE (1989) Choline-linked phosphoglycerides. A source of phosphatidic acid and diglycerides in stimulated neutrophils. J Biol Chem 264:1405-1413

Agwu DE, McPhail LC, Sozzani S, Bass DA, McCall CE (1991) Phosphatidic acid as a second messenger in human polymorphonuclear leukocytes. J Clin Invest 88:531-539

Ahmed S, Lee J, Kozma R, Best A, Monfries C, Lim L (1993) A novel functional target for tumor-promoting phorbol esters and lysophosphatidic acid. The p21[rac]-GTPase activating protein *n*-chimaerin. J Biol Chem 268:10709-10712

Aktories K, Braun U, Rosener S, Just I, Hall A (1989) The *rho* gene product expressed in *E.coli* is a substrate of botulinum ADP-ribosyltransferase C3. Biochem Biophys Res Commun 158:209-213

Al-Awar O, Radhakrishna H, Powell NN, Donaldson JG (2000) Separation of membrane trafficking and actin remodeling functions of ARF6 with an effector domain mutant. Molec Cell Biol 20:5998-6007

Andresen BT, Jackson EK, Romero GG (2001) Angiotensin II signaling to phospholipase D in renal microvascular smooth muscle cells in SHR. Hypertension 37:635-639

Arneson LS, Kunz J, Anderson RA, Traub LM (1999) Coupled inositide phosphorylation and phospholipase D activation initiates clathrin-coat assembly on lysosomes. J Biol Chem 274:17794-17805

Augert G, Bocckino SB, Blackmore PF, Exton JH (1989) Hormonal stimulation of diacylglycerol formation in hepatocytes. Evidence for phosphatidylcholine breakdown. J Biol Chem 264:21689-21698

Bae CD, Min DS, Fleming IN, Exton JH (1998) Determination of interaction sites on the small G protein RhoA for phospholipase D. J Biol Chem 273:11596-11604

Balboa MA, Firestein BL, Godson C, Bekk KS, Insel PA (1994) Protein kinase C_α mediates phospholipase D activation by nucleotides and phorbol ester in Madin-Darby canine kidney cells. Stimulation of phospholipase D is independent of activation of polyphosphoinositide-specific phospholipase C and phospholipase A_2. J Biol Chem 269:10511-10516

Baldassare JJ, Henderson PA, Burns D, Loomis C, Fisher GJ (1992) Translocation of protein kinase C isozymes in thrombin-stimulated human platelets. Correlation with 1,2-diacylglycerol levels. J Biol Chem 267:15585-15590

Bandoh K, Aoki J, Taira A, Tsujimoto M, Arai H, Inoue K (2000) Lysophosphatidic acid (LPA) receptors of the EDG family are differentially activated by LPA species. FEBS Lett 478: 159-165

Banno Y, Tamiya-Koizumi K, Oshima H, Morikawa A, Yoshida S, Nozawa Y (1997) Nuclear ADP-ribosylation factor (ARF)- and oleate-dependent phospholipase D (PLD) in rat liver cells. Increases of ARF-dependent PLD activity in regenerating liver cells. J Biol Chem 272:5208-5213

Banschbach MW, Geison RL, Hokin-Neaverson M (1981) Effects of cholinergic stimulation of levels and fatty acid composition of diacylglycerols in mouse pancreas. Biochim Biophys Acta 663:34-45

Bass DA, McPhail LC, Schmitt JD, Morris-Natschke S, McCall CE, Wykle RL (1989) Selective priming of rate and duration of the respiratory burst of neutrophils by 1,2-diacyl and 1-O-alkyl-2-acyl diglycerides. Possible relation to effects on protein kinase C. J Biol Chem 264:19610-19617

Bauldry SA,Bass DA, Cousart SL, McCall CE (1991) Tumor necrosis factor α priming of phospholipase D in human neutrophils. Correlation between phosphatidic acid production and superoxide generation. J Biol Chem 266:4173-4179

Bauldry SA, Elsey KL, Bass DA (1992) Activation of NADPH oxidase and phospholipase D in permeabilized human neutrophils. Correlation between oxidase activation and phosphatidic acid producton. J Biol Chem 267:25141-25152

Bellavite P, Corso F, Dusi S, Grzeskowiak M, Della-Bianca V, Rossi F (1988) Activation of NADPH-dependent superoxide production in plasma membrane extracts of pig neutrophils by phosphatidic acid. J Biol Chem 263:8210-8214

Benistant C, Rubin R (1990) Ethanol inhibits thrombin-induced secretion by human platelets at a site distinct from phospholipase C or protein kinase C. Biochem J 269:489-497

Bi K, Roth MG, Ktistakis NT (1997) Phosphatidic acid formation by phospholipase D is required for transport from the endoplasmic reticulum to the Golgi complex. Curr Biol 7:301-307

Billah MM, Pai J-K, Mullmann TJ, Egan RW, Siegel MI (1989) Regulation of phospholipase D in HL-60 granulocytes. Activation by phorbol esters, diglyceride, and calcium ionophore via protein kinase C-independent mechanisms. J Biol Chem 264:9069-9076

Bocckino SB, Blackmore PF, Exton JH (1985) Stimulation of 1,2-diacylglycerol accumulation in hepatocytes by vasopressin, epinephrine, and angiotensin II. J Biol Chem 260:14201-14207

Bocckino SB, Blackmore PF, Wilson PB, Exton JH (1987) Phosphatidate accumulation in hormone-treated hepatocytes via a phospholipase D mechanism. J Biol Chem 262:15309-15315

Bocckino SB, Wilson PB, Exton JH (1987) Ca^{2+}-mobilizing hormones elicit phosphatidylethanol accumulation via phospholipase D activation. FEBS Lett 225:201-204

Bocckino SB, Wilson PB, Exton JH (1991) Phosphatidate-dependent protein phosphorylation. Proc Natl Acad Sci USA 88:6210-6213

Boisgard R, Chanat E (2000) Phospholipase D-dependent and -independent mechanisms are involved in milk protein secretion in rabbit mammary epithelial cells. Biochim Biophys Acta 1495:281-296

Bokoch GM, Bohl BP, Chuang T-H (1994) Guanine nucleotide exchange regulates membrane translocation of Rac/Rho GTP-binding proteins. J Biol Chem 269:31674-31679

Bonser RW, Thompson NT, Randall RW, Garland LG (1989) Phospholipase D is functionally linked to superoxide generation in the human neutrophil. Biochem J 264:617-620

Bourgoin S, Grinstein S (1992) Peroxides of vanadate induce activation of phospholipase D in HL-60 cells. Role of tyrosine phosphorylation. J Biol Chem 267:11908-11916

Bourgoin S, Harbour D, Desmarais Y, Takai Y, Beaulieu A (1995) Low molecular weight GTP-binding proteins in HL-60 granulocytes. Assessment of the role of ARF and of a 50-kDa cytosolic protein in phospholipase D activation. J Biol Chem 270:3172-3178

Brinson AE, Harding T, Diliberto PA, He Y, Li X, Hunter D, Herman B, Earp HS, Graves LM (1998) Regulation of a calcium-dependent tyrosine kinase in vascular smooth muscle cells by angiotensin II and platelet-derived growth factor. Dependence on calcium and the actin cytoskeleton. J Biol Chem 273:1711-1718

Brisco CP, Martin A, Cross M, Wakelam MJO (1995) The roles of multiple pathways in regulating bombesin-stimulated phospholipase D activity in Swiss 3T3 fibroblasts. Biochem J306:115-122

Brown FD, Thomas N, Saqib KM, Clark JM, Powner D, Thompson NT, Solari R, Wakelam MJO (1998) Phospholipase D1 localises to secretory granules and lysosomes and is plasma-membrane translocated on cellular stimulation. Curr Biol 8:835-838

Brown HA, Gutowski S, Moomaw CR, Slaughter C, Sternweis PC (1993) ADP-ribosylation factor, a small GTP-dependent regulatory protein, stimulates phospholipase D activity. Cell 75:1137-1144

Brown HA, Gutowski S, Kahn RA, Sternweis PC (1995) Partial purification and characterization of Arf-sensitive phospholipase D from procine brain. J Biol Chem 270:14935-14943

Buhl AM, Johnson NL, Dhanasekaran N, Johnson GL (1995) $G\alpha_{12}$ and $G\alpha_{13}$ stimulate Rho-dependent stress fiber formation and focal adhesion assembly. J Biol Chem 270:14935-14943

Burnham DN, Uhlinger DJ, Lambeth JD (1990) Diradylglycerol synergizes with an anionic amphiphile to activate superoxide generation and phosphorylation of p47phox in a cell-free system from human neutrophils. J Biol Chem 265:17550-17559

Cabot MC, Jaken S (1984) Structural and chemical specificity of diradylglycerols for protein kinase C activation. Biochem Biophys Res Commun 125:163-169

Cai S, Exton JH (2001) Determination of interaction sites of phospholipase D_1 for RhoA. Biochem J 355:779-785

Carnero A, Cuadrado A, del Peso L, Lacal JC (1994) Activation of type D phospholipase by serum stimulation and ras-induced transformation in NIH3T3 cell. Oncogene 9:1387-1395

Carnero A, Dolfi F, Lacal JC (1994) ras-p21 activates phospholipase D and A2, but not phospholipase C or PKC, in Xenopus laevis oocytes. J Cell Biochem 54:478-486

Caumont A-S, Galas M-C, Vitale N, Aunis D, Bader M-F (1998) Regulated exocytosis in chromaffin cells. Translocation of ARF6 stimulates a plasma membrane-associated phospholipase D. J Biol Chem 273:1373-1379

Caumont A-S, Vitale N, Gensse M, Galas, M-C, Casanova JE, Bader M-F (2000) Identification of a plasma membrane-associated guanine nucleotide exchange factor for ARF6 in chromaffin cells. Possible role in the regulated exocytotic pathway. J Biol Chem 275: 15637-15644

Cavenagh MM, Whitney JA, Carroll K, Zhang C-j, Bowman AL, Rosenwald AG, Mellman I, Kahn RA (1996) Intracellular distribution of Arf proteins in mammalian cells. Arf6 is uniquely localized to the plasma membrane. J Biol Chem 271:21767-21774

Chao W, Liu H, Hanahan DJ, Olson MS (1992) Platelet-activating factor-stimulated protein tyrosine phosphorylation and eicosanoid synthesis in rat kupffer cells. Evidence for calcium-dependent and protein kinase C-dependent and -independent pathways. J Biol Chem 267: 6725-6735

Chen Y-G, Siddhanta A, Austin CD, Hammond SM, Sung T-C, Frohman MA, Morris AJ, Shields D (1997) Phospholipase D stimulates release of nascent secretory vesicles from the trans-Golgi network. J Cell Biol 138:495-504

Ching T-T, Wang D-S, Hsu A-L, Lu P-J, Chen C-S (1999) Identification of multiple-specific phospholipases D as new regulatory enzymes for phosphatidylinositol 3,4,5-trisphosphate. J Biol Chem 274:8611-88617

Chong LD, Traynor-Kaplan A, Bokoch GM, Schwartz MA (1994) The small GTP-binding protein Rho regulates a phosphatidylinositol 4-phosphate 5-kinase in mammalian cells. Cell 79:507-513

Chuang T-H, Bohl BB, Bokoch GM (1993) Biologically active lipids are regulators of Rac GDI complexation. J Biol Chem 268:26206-26211

Chung J-K, Sekiya F, Kang H-S, Lee C, Han J-S, Kim SR, Bae YS, Morris AJ, Rhee SG (1997) Synaptojanin inhibition of phospholipase D activity by hydrolysis of phosphatidylinositol 4,5-bisphosphate. J Biol Chem 272:15980-15985

Cockcroft S, Allan D (1984) The fatty acid composition of phosphatidylinositol, phosphatidate and 1,2-diacylglycerol in stimulated human neutrophils. Biochem J 222:557-559

Cockcroft S, Thomas GMH, Fensome A, Geny B, Cunningham E, Gout I, Hiles I, Totty NF, Truong O, Hsuan JJ (1994) Phospholipase D: A downstream effector of ARF in granulocytes. Science 263:523-526

Colley WC, Sung T-C, Roll R, Jenco J, Hammond SM, Altshuller Y, Bar-Sagi D, Morris AJ, Frohman MA (1997) Phospholipase D2, a distinct phospholipase D isoform with novel regulatory properties that provokes cytoskeletal reorganization. Curr Biol 7:191-201

Conricode KM, Brewer KA, Exton JH (1992) Activation of phospholipase D by protein kinase C. Evidence for a phosphorylation-independent mechanism. J Biol Chem 267:7199-7202

Conricode KM, Smith JL, Burns DJ, Exton JH (1994) Phospholipase D activation in fibroblast membranes by the (and (isoforms of protein kinase C. FEBS Lett. 342:149-153

Contos JJA, Ishii I, Chun J (2000) Lysophosphatidic acid receptors. Molec Pharmacol 58:1188-1196

Cross MJ, Roberts S, Ridley AJ, Hodgkin MN, Stewart A, Claesson-Welsh L,Wakelam MJO (1996) Stimulation of actin stress fibre formation mediated by activation of phospholipase D. Curr Biol 6:588-597

Czarny M, Lavie Y, Fiucci G, Liscovitch M (1999) Localization of phospholipase D in detergent-insoluble, caveolin-rich membrane domains. Modulation by caveolin-1expression and caveolin-1$_{182-101}$. J Biol Chem 274:2717-2724

Czarny M, Fiucci G, Lavie Y, Banno Y, Nozawa Y, Liscovitch M (2000) Phospholipase D2: Functional interaction with caveolin in low density membrane microdomains. FEBS Letts 467:326-332

Daniel LW, Small GW, Schmitt JD (1988) Alkyl-linked diglycerides inhibit protein kinase C activation by diacylglycerols. Biochem Biophys Res Commun 151:291-297

Dennis EA (1994) Diversity of group types, regulation, and function of phospholipase A$_2$. J Biol Chem 269:13057-13061

Divecha N, Lander DJ, Scott TW, Irvine RF (1991) Molecular species analysis of 1,2-diacylglycerols and phosphatidic acid formed during bombesin stimulation of Swiss 3T3 cells. Biochim Biophys Acta 1093:184-188

Divecha N, Roefs M, Halstead JR, D'Andrea S, Fernandez-Borga M, Oomen L, Saqib KM, Wakelam MJO, D'Santos C (2000) Interaction of the type I(PIP kinase with phospholipase D: a role for the local generation of phosphatidylinositol 4,5-bisphosphate in the regulation of PLD2 activity. EMBO J 19:5440-5449

Donaldson JG, Lippincott-Schwartz J (2000) Sorting and signaling at the Golgi complex. Cell 101:693-696

Dougherty RW, Dubay GR, Niedel JE (1989) Dynamics of the diradylglycerol responses of stimulated phagocytes. J Biol Chem 264:11263-11269

Du G, Altshuller YM, Kim Y, Han JM, Ryu SH, Morris AJ, Frohman MA, Wakelam MJO, D'Santos C (2000a) Interaction of the type I(PIP kinase with phospholipase D: a role for the local generation of phosphatidylinositol 4,5-bisphosphate in the regulation of PLD2 activity. EMBO J 19:5440-5449

Du G, Altshuller YM, Kim Y, Han JM, Ryu SH, Morris AJ, Frohman MA (2000b) Dual requirement for Rho and protein kinase C in direct activation of phospholipase D1 through G protein-coupled receptor signaling. Molec Biol Cell 11:4359-4368

Dubyak GR, Schomisch SJ, Kusner DJ, Xie M (1993) Phospholipase D activity in phagocytic leucocytes is synergistically regulated by G-protein- and tyrosine kinase-based mechanisms. Biochem J 292:121-128

Eichholtz T, Jalink K, Fahrenfort I, Moolenaar WH (1993) The bioactive phospholipid lysophosphatidic acid is released from activated platelets. Biochem J 291:677-680

El Bawab S, Macovschi O, Sette C, Conti M, Lagarde M, Nemoz G, Prigent A-F (1997) Selective stimulation of a cAMP-specific phosphodiesterase (PDE4A5) isoform by phosphatidic acid molecular species endogenously formed in rat thymocytes. Eur J Biochem 247:1151-1157

Eldar H, Ben-AV P, Schmidt U-S, Livneh E, Liscovitch M (1993) Upregulation of phospholipase D activity induced by overexpression of protein kinase C-α. Studies in intact Swiss/3T3 cells and in detergent-solubilized membranes in vitro. J. Biol. Chem. 268:12560-12564

El Hadj NB, Popoff MR, Marvaud J-C, Payrastre B, Boquet P, Geny B (1999) G-protein-stimulated phospholipase D activity is inhibited by lethal toxin from *Clostridium sordellii* in HL-60 cells. J Biol Chem 274:14021-14031

Emoto M, Klarlund JK, Waters SB, Hu V, Buxton JM, Chawla A, Czech MP (2000) A role for phospholipase D in GLUT4 glucose transporter translocation. J Biol Chem 275:7144-7151

Epand RM, Stafford AR (1990) Counter-regulatory effects of phosphatidic acid on protein kinase C activity in the presence of calcium and diolein. Biochem Biophys Res Commun 171:487-490

Erickson RW, Langel-Peveri P, Traynor-Kaplan AE, Heyworth PG, Curnutte JT (1999) Activation of human neutrophil NADPH oxidase by phosphatidic acid or diacylglycerol in a cell-free system. Activity of diacylglycerol is dependent on its conversion to phosphatidic acid. J Biol Chem 274:22243-22250

Exton JH (1990) Signaling through phosphatidylcholine breakdown. J Biol Chem 265:1-4

Exton JH (1996) Regulation of phosphoinositide phospholipases by hormones, neurotransmitters, and other agonists linked to G proteins. Annu. Rev. Pharmacol.Toxicol. 36:481-509

Exton JH (1997) Phospholipase D: Enzymology, mechanisms of regulation, and function. Physiol Rev 77:303-320

Exton JH (1998) Phospholipase D. Biochim Biophys Acta 1436:105-115

Exton JH (1999) Regulation of phospholipase D. Biochim Biophys Acta 1439:121-133

Fällman M, Gullberg M, Hellberg C, Andersson T (1992) Complement receptor-mediated phagocytosis is associated with accumulation of phosphatidylcholine-derived digylceride in human neutrophils. Involvement of phospholipase D and direct evidence for a positive feedback signal of protein kinase C. J Biol Chem 267:2656-2663

Fang Y, Vilella-Bach M, Bachmann R, Flanigan A, Chen J (2001) Phosphatidic acid mediated mitogenic activation of mTOR signaling. Science In press.

Fensome A, Cunningham E, Prosser S, Tan SK, Swigart P, Thomas G, Hsuan J, Cockcroft S (1996) ARF and PITP restore GTP(S-stimulated protein secretion from cytosol-depleted HL60 cells by promoting PIP2 synthesis. Curr Biol 6:730-738

Fensome A, Whatmore J, Morgan C, Jones D, Cockcroft S (1998) ADP-ribosylation factor and Rho proteins mediate fMLP-dependent activation of phospholipase D in human neutrophils. J Biol Chem 273:13157-13164

Fleming I, Elliott CM, Exton JH (1996) Differential translocation of Rho family GTPases by lysophosphatidic acid, endothelin-1,and platelet-derived growth factor. J Biol Chem 271:33067-33073

Frank C, Keilhack H. Opitz F, Zschörnig O, Böhmer F-D (1999) Binding of phosphatidic acid to the protein-tyrosine phosphatase SHP-1 as a basis for activity modulation. Biochem 38:11993-12002

Frankel P, Ramos M, Flom J, Bychenok S, Joseph T, Kerkhoff E, Rapp UR, Feig LA, Foster DA (1999) Ral and Rho-dependent activation of phospholipase D in v-Raf-transformed cells. Biochem Biophys Res Commun 255:502-507

Freeman EJ (2000) The Ang II-induced growth of vascular smooth muscle cells involves a phospholipase D-mediated signaling mechanism. Arch Biochem Biophys 374:363-370

Freyberg Z, Sweeney D, Siddhanta A, Bourgoin S, Frohman M, Shields D (2001) Intracellular localization of phospholipase D1 in mammalian cells. Mol Biol Cell 12:943-955

Frohman M, Sung T-C, Morris AJ (1999) Mammalian phospholipase D structure and regulation. Biochim Biophys Acta 1439:175-186

Fromm C, Coso OA, Montaner S, Xu N, Gutkind JS (1997) The small GTP-binding protein Rho links G protein-coupled receptors and $G\alpha_{12}$ to the serum response element and to cellular transformation. Proc Natl Acad Sci USA 94:10098-10103

Fukami K, Takenawa T (1992) Phosphatidic acid that accumulates in platelet-derived growth factor-stimulated Balb/c 3T3 cells is a potential mitogenic signal. J Biol Chem 267:10988-10993

Ganley IG, Walker, SJ, Manifava M, Li D, Brown A, Ktistakis NT (2001) Interaction of phospholipase D1 with a casein-kinase-2-like serine kinase. Biochem J 354:369-378

Ganong BR, Loomis CR, Hannun YA, Bell RM (1986) Specificity and mechanism of protein kinase C activation by sn-1,2-diacylglycerols. Proc Natl Acad Sci USA 83:1184-1188

Gaschet J, Hsu VW (1999) Distribution of ARF6 between membrane and cytosol is regulated by its GTPase cycle. J Biol Chem 274:20040-20045

Ghosh S, Strum JC, Sciorra VA, Daniel L, Bell RM (1996) Raf-1 kinase possesses distinct binding domains for phosphatidylserine and phosphatidic acid. Phosphatidic acid regulates the translocation of raf-1 in 12-0-tetradecanoylphorbol-13-acetate-stimulated madin-darby canine kidney cells. J Biol Chem 271:8472-8480

Gillooly DJ, Melendez AJ, Hockaday AR, Hanett MM, Allen JM (1999) Endocytosis and vesicular trafficking of immune complexes and activation of phospholipase D by the human high-affinity IgG receptor requires distinct phosphoinositide 3-kinase activities. Biochem J 344:605-611

Go M, Sekiguchi K, Nomura H, Kikkawa U, Nishizuka Y (1987) Further studies on the specificity of diacylglycerol for protein kinase C activation. Biochem Biophys Res Commun 144:598-605

Godi A, Pertile P, Meyers R, Marra P, Di Tullio G, Iurisci C, Luini A, Corda D, De Matteis MA (1999) ARF mediates recruitment of PtdIns-4-OH kinase-β and stimulates synthesis of PtdIns $(4,5)P_2$ on the Golgi complex. Nat Cell Biol 1:280-287

Gomez-Cambronero J (1995) Immunoprecipitation of a phospholipase D activity with antiphosphotyrosine antibodies. J Inter Cyto Res 15:877-885

Gómez-Muñoz A, Martin A, O'Brien L, Brindley, DN (1994) Cell-permeable ceramides inhibit the stimulation of DNA synthesis and phospholipase D activity by phosphatidate and lysophosphatidate in rat fibroblasts. J Biol Chem 269:8937-8943

Gómez-Muñoz A, Waggoner DW, O'Brien L, Brindley DN (1995) Interaction of ceramides, sphingosine, and sphingosine 1-phosphate in regulating DNA synthesis and phospholipase D activity. J Biol Chem 270:26318-26325

Gottlin EB, Rudolph AE, Zhao Y, Matthews HR, Dixon JE (1998) Catalytic mechanism of the phospholipase D superfamily proceeds via a covalent phosphohistidine intermediate. Proc Natl Acad Sci USA 95:9202-9207

Grange M, Sette C, Cuomo M, Conti M, Lagarde M, Prigent A-F, Némoz G (2000) The cAMP-specific phosphodiesterase PDE4D3 is regulated by phosphatidic acid binding. Consequences for cAMP signaling pathway and characterization of a phosphatidic acid binding site. J Biol Chem 275:33379-33387

Gruchalla RS, Dinh TT, Kennerly DA (1990) An indirect pathway of receptor-mediated 1,2-diacylglycerol formation in mast cells. I. IgE receptor-mediated activation of phospholipase D. J Immunol 144:2334-2342

Gu F, Gruenberg J (2000) ARF1 regulates pH-dependent COP functions in the early endocytic pathway. J Biol Chem 275:8154-8160

Guillemain I, Exton JH (1997) Effects of brefeldin A on phosphatidylcholine phospholipase D and inositolphospholipid metabolism in HL-60 cells. Eur J Biochem 249:812-819

Gustavsson L, Moehren, G, Torres-Marquez ME, Benistant C, Rubin R, Hoek JB (1994) The role of cytosolic Ca^{2+}, protein kinase C, and protein kinase A in hormonal stimulation of phospholipase D in rat hepatocytes. J Biol Chem 269:849-858

Ha K-S, Exton JH (1993b) Activation of actin polymerization by phosphatidic acid derived from phosphatidylcholine in IIC9 fibroblasts. J Cell Biol 123:1789-1796

Ha K-S, Exton JH (1993bb) Differential translocation of protein kinase C isozymes by thrombin and platelet-derived growth factor. A possible function for phosphatidylcholine-derived diacylglycerol. J Biol Chem 268:10534-10539

Ha K-S, Yeo E-J, Exton JH (1994) Lysophosphatidic acid activation of phosphatidylcholine-hydrolysing phospholipase D and actin polymerization by a pertussis toxin-sensitive mechanism. Biochem J 303:55-59

Hammond SM, Altshuller YM, Sung T-C, Rudge SA, Rose K, Engebrecht JA, Morris AJ, Frohman MA (1995) Human ADP-ribosylation factor-activated phosphatidylcholine-specific phospholipase D defines a new and highly conserved gene family. J Biol Chem 270:29640-29643

Hammond SM, Jenco JM, Nakashima S, Cadwallader K, Gu Q-m, Cook S, Nozawa Y, Prestwich GD, Frohman MA, Morris AJ (1997) Characterization of two alternately spliced forms of phospholipase D1. Activation of the purified enzymes by phosphatidylinositol 4,5-bisphosphate, ADP-ribosylation factor, and Rho family monomeric GTP-binding proteins and protein kinase C-α. J Biol Chem 272:3860-3868

Han J-S, Chung J-K, Kang H-S, Donaldson J, Bae YS, Rhee SG (1996) Multiple forms of phospholipase D inhibitor from rat brain cytosol. Purification and characterization of heat-labile form. J Biol Chem 271:11163-11169

Hannun YA, Loomis CR, Bell RM(1986) Protein kinase C activated in mixed micelles. Mechanistic implications of phospholipid, diacylglycerol, and calcium interdependencies. J Biol Chem 261:7184-7190

Hardwick JS, Sefton BM (1997) The activated form of the Lck tyrosine protein kinase in cells exposed to hydrogen peroxide is phosphorylated at both Tyr-394 and Tyr-505. J Biol Chem 272:25429-25432

Hermans SWG, Engelmann B, Reinhardt U, Bartholomeus-Van Nooij IGP, De Pont JJHHM, Willems PHGM (1996) Diradylglycerol formation in cholecystokinin-stimulated rabbit pancreatic acini assessment of precursor phospholipids by means of molecular species analysis. Eur J Biochem 235:73-81

Hess JA, Ross AH, Qiu R-G, Symons M, Exton JH (1997) Role of Rho family proteins in phospholipase D activation by growth factors. J Biol Chem 272:1615-1620

Hodgkin MN, Clark JM, Rose S, Saqib K, Wakelam MJO (1999) Characterization of the regulation of phospholipase D activity in the detergent-insoluble fraction of HL60 cells by protein kinase C and small G-proteins. Biochem J 339:87-93

Hodgkin MN, Masson MR, Powner D, Saqib KM, Ponting CP, Wakelam MJO (1999) Phospholipase D regulation and localization is dependent upon a phosphatidyli-nositol 4,5-bisphosphate-specific PH domain. Curr Biol 10:43-46

Hoer A, Cetindag C, Oberdisse E (2000) Influence of phosphatidylinositol 4,5-bisphosphate on human phospholipase D1 wild-type and deletion mutants: is there evidence for an interaction of phosphatidylinositol 4,5-bisphosphate with the putative pleckstrin homology domain. Biochim Biophys Acta 1481:189-201

Holbrook PG, Pannell LK, Daly JW (1991) Phospholipase D-catalyzed hydrolysis of phosphatidylcholine occurs with P-O bond cleavage. Biochim Biophys Acta 1984:155-158

Holub BJ, Kuksis A (1978) Metabolism of molecular species of diacylglycerophos-pholipids. Adv Lipid Res 16:1-125

Honda A, Nogami M, Yokozeki T, Yamazaki M, Nakamura H, Watanabe H, Ka-wamoto K, Nakayama K, Morris AJ, Frohman MA, Kanaho Y (1999) Phosphati-dylinositol 4-phosphate 5-kinase α is a downstream effector of the small G protein ARF6 in membrane ruffle formation. Cell 99:521-532

Hong J-H, Oh S-O, Lee M, Kim Y-R, Kim D-U, Hur GM, Lee JH, Lim K, Hwang B-D, Park S-K (2001) Enhancement of lysophosphatidic acid-induced ERK phosphory-lation by phospholipase D1 via the formation of phosphatidic acid. Biochem Biophys Res Commun 281:1337-1342

Hong J-X, Lee F-JS, Patton WA, Lin CY, Moss J, Vaughan M (1998) Phospholipid- and GTP-dependent activation of cholera toxin and phospholipase D by human ADP-ribosylation factor-like protein 1 (HARL1). J Biol Chem 273:15872-15876

Hooley R, Yu C-Y, Symons M, Barber DL (1996) $G\alpha13$ stimulates Na^+-H^+ exchange through distinct Cdc42-dependent and RhoA-dependent pathways. J Biol Chem 271:6152-6168

Horn JM, Lehman JA, Alter G, Horwitz J, Gomez-Cambronero J (2001) Presence of a phospholipase D (PLD) distinct from PLD1 or PLD2 in human neutrophils: im-munobiochemical characterization and initial purification. Biochim Biophys Acta 1530:97-110

Houle MG, Kahn RA, Naccache PH, Bourgoin S (1995) ADP-ribosylation factor translocation correlates with potentiation of GTPγS-stimulated phospholipase D activity in membrane fractions of HL-60 cells. J Biol Chem 270:22795-22800

Houle MG, Naccache PH, Bourgoin S (1999) Tyrosine kinase-regulated small GTPase translocation and the activation of phospholipase D in HL60 granulocytes. J Leu-koc Biol 66:1021-1030

Huang C, Cabot MC (1990) Vasopressin-induced polyphospoinositide and phos-phatidylcholine degradation in fibroblasts. Temporal relationship for formation of phospholipase C and phospholipase D hydrolysis products. J Biol Chem 265:17468-17473

Huang C-K, Bonak V, Laramee GR, Casnellie JE (1990) Protein tyrosine phosphorylation in rabbit peritoneal neutrophils. Biochem J 269:431-436

Huckle WR, Dy RC, Earp HS (1992) Calcium-dependent increase in tyrosine kinase activity stimulated by angiotensin II. Proc Natl Acad Sci USA 89:8837-8841

Hurst KM, Chataway TK, Hughes BP, Barritt GJ (1991) Low molecular weight GTP-binding proteins in hepatocytes and an assessment of the role of p21^{ras} proteins in the activation of phospholipase D. Biochem Internat 24:507-516

Irving HR, Exton JH (1987) Phosphatidylcholine breakdown in rat liver plasma membranes. Roles of guanine nucleotides and P2-purinergic agonists. J Biol Chem 262:3440-3443

Ito Y, Nakashima S, Kanoh H, Nozawa Y (1997a) Implication of Ca^{2+}-dependent protein Tyrosine phosphorylation in carbachol-induced phospholipase D activation in rat Pheochromocytoma PC12 cells. J Neurochem 68:419-425

Ito Y, Nakashima S, Nozawa Y (1997b) Hydrogen peroxide-induced phospholipase D activation in rat pheochromocytoma PC12 cells: Possible involvement of Ca^{2+}-dependent protein tyrosine kinase. J Neurochem 69:729-736

Iyer SS, Kusner DJ (1999) Association of phospholipase D activity with the detergent-insoluble cytoskeleton of I937 promonocytic leukocytes. J Biol Chem 274:2350-2359

Jackowski S, Rock CO (1989) Stimulation of phosphatidylinositol 4,5-bisphosphate phospholipase C activity by phosphatidic acid. Arch Biochem Biophys 268:516-524

Jaken S (1997 Protein kinase C intracellular binding proteins. In: Molecular Biology Intelligence Unit. Eds.: Parker PJ & Dekker LV, R.G. Landes Company (Austin), pp179-188

Jenco JM, Rawlingson A, Daniels B, Morris AJ (1998) Regulation of phospholipase D2: Selective inhibition of mammalian phospholipase D isoenzymes by α- and β-synucleins. Biochem 37:4901-4909

Jenkins GH, Fisette PL, Anderson RA (1994) Type I phosphatidylinositol 4-phosphate 5-kinase isoforms are specifically stimulated by phosphatidic acid. J Biol Chem 269:11547-11554

Jiang H, Lu Z, Luo J-Q, Wolfman A, Foster DA (1995a) Ras mediates the activation of phospholipase D by v-Src. J Biol Chem 270:6006-6009

Jiang H, Luo J-Q, Urano T, Frankel P, Lu Z, Foster DA, Feig LA (1995b) Involvement of Ral GTPase in v-Src-induced phospholipase D activation. Nature 378:409-412

Jinsi A, Paradise J, Deth RC (1996) A tyrosine kinase regulates α-adrenoceptor-stimulated contraction and phospholipase D activation in the rat aorta. Eur J Pharmacol 302:183-190

Jones AT, Clague MJ (1997) Regulation of early endosome fusion by phospholipase D activity. Biochem Biophys Res Commun 236:285-288

Jones DH, Bax B, Fensome A, Cockcroft S (1999) ADP ribosylation factor 1 mutants identify a phospholipase D effector region and reveal that phospholipase D participates in lysosomal secretion but is not sufficient for recruitment of coatomer 1. Biochem J 341:185-192

Jones DH, Morris JB, Morgan CP, Kondo H, Irvine RF, Cockcroft S (2000) Type I phosphatidylinositol 4-phosphate 5-kinase directly interacts with ADP-ribosylation factor 1 and is responsible for phosphatidylinositol 4,5-bisphosphate synthesis in the Golgi compartment. J Biol Chem 275:13962-13966

Jones GA, Carpenter G (1993) The regulation of phospholipase C-γ1 by phosphatidic acid. Assessment of kinetic parameters. J Biol Chem 268:20845-20850

Jones MJ, Murray AW (1995) Evidence that ceramide selectively inhibits protein kinase C-α translocation and modulates bradykinin activation of phospholipase D. J Biol Chem 270:5007-5013

Just I, Selzer J, Wilm M, von Eichel-Streiber C, Mann M, Aktories K (1995) Glucosylation of Rho proteins by *Clostridium difficile* toxin B. Nature 375:500-503

Kahn RA, Yucel JK, Malhotra V (1993) ARF signaling: A potential role for phospholipase D in membrane traffic. Cell 75:1045-1048

Khan WA, Blobe GC, Richards AL, Hanun YA (1994) Identification, partial purification, and characterization of a novel phospholipid-dependent and fatty acid-activated protein kinase from human platelets. J Biol Chem 269:9729-9735

Kam Y, Exton JH (2001) Phospholipase D activity is required for actin stress fiber formation in fibroblasts. Molec Cell Biol 21:4055-4066

Kanoh H, Williger B-T, Exton JH (1997) Arfaptin 1, a putative cytosolic target protein of ADP-ribosylation facator, is recruited to Golgi membranes. J Biol Chem 272:5421-5429

Katayama K, Kodaki T, Nagamachi Y, Yamashita S (1998) Cloning, differential regulation and tissue distribution of alternatively spliced isoforms of ADP-ribosylation-factor-dependent phospholipase D from rat liver. Biochem J 329:647-652

Katoh H, Aoki J, Yamaguchi Y, Kitano Y, Ichikawa A, Negishi M (1998) Constitutively active $G\alpha_{12}$, $G\alpha_{13}$, and $G\alpha_q$ induce Rho-dependent neurite retraction through different signaling pathways. J Biol Chem 273:28700-28707

Kaszkin M, Richards J, Kinzel V (1992) Proposed role of phosphatidic acid in the extracellular control of the transition from G2 to mitosis exerted by epidermal growth factor in A431 cells. Canc Res 52627-5634

Keeley SJ, Calandrella SO, Barrett KE (2000) Carbachol-stimulated transactivation of epidermal growth factor receptor and mitogen-activated protein kinase in T84 cells is mediated by intracellular Ca^{2+}, PYK-2, and p60[src]. J Biol Chem 275:12619-12625

Keller J, Schmidt M, Hussein B, Rumenapp U, Jakobs KH (1997) Muscarinic receptor-stimulated cytosol-membrane translocation of RhoA. FEBS Lett 403:299-302

Kennerly D (1990) Phosphatidylcholine is a quantitatively more important source of increased 1,2-diacylglycerol than is phosphatidylinositol in mast cells. J Immunol 144:3912-3919

Kessels GCR, Roos D, Verhoeven AJ (1991) fMet-Leu-Phe-induced activation of phospholipase D in human neutrophils. Dependence on changes in cytosolic free Ca^{2+} concentration and relation with respiratory burst activation. J Biol Chem 266:23152-23156

Kiley S, Schapp D, Parker P, Hsieh L-L, Jaken S (1990) Protein kinase C heterogeneity in GH_4C_1 rat pituitary cells. J Biol Chem 265:15704-15712

Kim JH, Lee SD, Han JM, Lee TG, Kim Y, Park JB, Lambeth JD, Suh P-G, Ryu SH (1998) Activation of phospholipase D1 by direct interaction with ADP-ribosylation factor 1 and RalA. FEBS Lett 430:231-235

Kim JH, Han JM, Lee S, Kim Y, Lee TG, Park JB, Lee SD, Suh P-G, Ryu SH (1999) Phospholipase D1 in caveolae: Regulation by protein kinase Cα and caveolin-1. Biochem J 38:3763-3769

Kim JH, Kim Y, Lee SD, Lopez I, Arnold RS, Lambeth JD, Suh P-G, Ryu SH (1999a) Selective activation of phospholipase D2 by unsaturated fatty acid. FEBS Lett 454:42-46

Kim Y, Kim J-E, Lee SD, Lee TG, Kim JH, Park JB, Han JM, Jang SK, Suh P-G, Ryu SH (1999b) Phospholipase D1 is located and activated by protein kinase Cα in the plasma membrane in 3Y1 fibroblast cell. Biochim Biophys Acta 1436:319-330

Kim Y, Han JM, Han BR, Lee K-A, Kim JH, Lee BD, Jang I-H, Suy P-G, Ryu SH (2000) Phospholipase D1 is phosphorylated and activated by protein kinase C in caveolin-enriched microdomains within the plasma membrane. J Biol Chem 275:13621-13627

Kishikawa K, Chalfant CE, Perry DK, Bielawska A, Hannun YA (1999) Phosphatidic acid is a potent and selective inhibitor of protein phosphatase 1 and an inhibitor of ceramide-mediated responses. J Biol Chem 274:21335-21341

Kiss Z, Petrovics G, Olah Z, Lehel C, Anderson WB (1999) Overexpression of protein kinase C-ε and its regulatory domains in fibroblasts inhibits phorbol ester-induced phospholipase D activity. Arch Biochem Biophys 363:121-128

Klein J, Chalifa V, Liscovitch M, Löffelholz K (1995) Role of phospholipase D activation in nervous system physiology and pathophysiology. J Neurochem 65:1445-1455

Kodaki T, Yamashita S (1997) Cloning, expression, and characterization of a novel phospholipase D complementary DNA from rat brain. J Biol Chem 272:11408-11413

Kondo T, Inui H, Konishi F, Inagami T (1992) Phospholipase D mimics platelet-derived growth factor as a competence factor in vascular smooth muscle cells. J Biol Chem 267:23609-23616

Koonin EV (1996) A duplicated catalytic motif in a new superfamily of phosphohydrolases and phospholipid synthases that includes poxvirus envelope proteins. Trends Biochem Sci 21:242-243

Kozawa O, Blume-Jensen P, Heldin C-H, Ronnstrand LL (1997) Involvement of phosphatidylinositol 3'-kinase in stem-cell-factor-induced phospholipase D activation and arachidonic acid release. Eur J Biochem 248:149-155

Kranenburg O, Moolenaar WH (2001) Ras-MAP kinase signaling by lysophosphatidic acid and other G protein-coupled receptor agonists. Oncogene 20:1540-1546

Kranenburg O, Poland M, Gebbink M, Oomen L, Moolenaar WH (1997) Dissociation of LPA-induced cytoskeletal contraction from stress fiber formation by differential localization of RhoA. J Cell Sci 110:2417-2427

Ktistakis NT, Brown HA, Sternweis PC, Roth MG (1995) Phospholipase D is present in Golgi -enriched membranes and its activation by ADP-ribosylation factor is sensitive to brefeldin A Proc Natl Acad Sci USA 92:4952-4956

Ktistakis NT, Brown HA, Waters, MG, Sternweis PC, Roth MG (1996) Evidence that phospholipase D mediates ADP ribosylation factor-dependent formation of Golgi coated vesicles. J Cell Biol 134:295-306

Kuai J, Boman AL, Arnold RS, Zhu X, Kahn RA (2000) Effects of activated ADP-ribosylation factors on Golgi morphology require neither activation of phospholipase D1 nor recruitment of coatomer. J Biol Chem 275:4022-4032

Kuksis A, Marai L, Breckendrige WC, Gornall DA, Stachnyk O (1968) Molecular species of lecithins of some functionally distinct rat tissues. Can J Physiol Pharmacol 46:511-524

Kumada T, Nakashima S, Nakamura Y, Miyata H, Nozawa Y (1995) Antigen-mediated phospholipase D activation in rat basophilic leukemia (RBL-2H3) cells: possible involvement of calcium/calmodulin. Biochim Biophys Acta 1258:107-114

Kuribara H, Tago K, Yokozeki T, Sasaki T,Takai Y, Morii N, Narumiya S, Katadda T, Kanaho Y (1995) Synergistic activation of rat brain phospholipase D by ADP-ribosylation factor and *rhoA* p21,and its inhibition by *Clostridium botulinum* C3 exoenzyme. J Biol Chem 270:25667-25671

Kusaka I, Ishikawa S-E, Higashiyama M, Saito T, Nagasaka S, Saito T (1996) The activation of phospholipase D participates in the mitogenic action of arginine vasopressin in cultured rat glomerular mesangial cells. Endocrinology 137:5421-5428

Kusner DJ, Schomisch SJ, Dubyak GR (1993) ATP-induced potentiation of G-protein-dependent phospholipase D activity in a cell-free system from U937 promonocytic leukocytes J Biol Chem 268:19973-19982

Leach KL, Ruff VA, Wright TM, Pessin MS, Raben DM (1991) Dissociation of protein kinase C activation and *sn*-1,2-diacylglycerol formation. Comparison of phosphatidylinositol- and phosphatidylcholine-derived diglycerides in α-thrombin-stimulated fibroblasts. J Biol Chem 265:3215-3221

Leduc I, Meloche S (1995) Angiotensin II stimulates tyrosine phosphorylation of the focal adhesion-associated protein paxillin in aortic smooth muscle cells. J Biol Chem 270:4401-4404

Lee C, Fisher SK, Agranoff BW, Hajra AK (1991) Quantitative analysis of molecular species of diacylglycerol and phosphatidate formed upon muscarinic receptor activation of human SK-N-SH neuroblastoma cells. J Biol Chem 266:22837-22846

Lee C, Kang H-S, Chung J-K, Sekiya F, Kim J-R, Han J-S, Kim SR, Bae YS, Morris AR, Rhee SG (1997) Inhibition of phospholipase D by clathrin assembly protein 3 (AP3). J Biol Chem 272:15986-15992

Lee S, Park JB, Kim JH, Kim Y, Kim JH, Kim Y, Kim JH, Shin K-J, Lee JS, Ha SH, Suh P-G, Ryu SH (2001) Actin directly interacts with phospholipase D inhibiting its activity. J Biol Chem 276:28252-28260

Lee SY, Yeo E-J, Choi M-U (1998) Phospholipase D activity in L1210 cells: A model for oleate-activated phospholipase D in intact mammalian cells. Biochem Biophys Res Commun 244:825-831

Lee YH, Kim HS, Pai J-K, Ryu SH, Suh P-G (1994) Activation of phospholipase D induced by platelet-derived growth factor is dependent upon the level of phospholipase C-γ1. J Biol Chem 269:26842-26847

Leiros I, Secundo F, Zambonelli C, Servi S, Hough E (2000) The first crystal structure of a phospholipase D. Structure 8:655-667

Le Stunff HL, Dokhac L, Bourgoin S, Bader M-F, Harbon S (2000a) Phospholipase D in rat myometrium: occurrence of a membrane-bound ARF6 (ADP-ribosylation factor 6)-regulated activity controlled by ((subunits of heterotrimeric G-proteins. Biochem J 352:491-499

Le Stunff H, Dokhac L, Harbon S (2000b) The roles of protein kinase C and tyrosine kinases in mediating endothelin-1-stimulated phospholipase D activity in rat myometrium: Differential inhibition by ceramides and cyclic AMP. J Pharmacol Exper Therap 292:628-637

Levade T, Jaffrézou J-P (1999) Signalling sphingomyelineases: which, where, how and why? Biochim Biophys Acta 1438:1-17

Li Y, Shiels AJ, Maszak G, ByronKL (2001) Vasopressin-stimulated Ca^{2+} spiking in vascular smooth muscle cells involves phospholipase D. Am J Physiol Heart Circ Physiol 280:H2658-H2664

Limatola C, Schaap D, Moolenaar WH, van Blitterswijk WJ (1994) Phosphatidic acid activation of protein kinase C-ζ overexpressed in COS cells: comparison with other protein kinase C isotypes and other acidic lipids. Biochem J 304:1001-1008

Lin P, Wiggan GA, Gilfillan AM (1991) Activation of phospholipase D in a rat mast (RBL 2H3) cell line. A possible unifying mechanism for IgE-dependent degranulation and arachidonic acid metabolite release. J Immunol 146:1609-1616

Liscovitch M, Chalifa V, Pertile P, Chen C-S, Cantley LC (1994) Novel function of phosphatidylinositol 4,5-bisphosphate as a cofactor for brain membrane phospholipase D. J Biol Chem 269:21403-21406

Liscovitch M, Czarny M, Fiucci G, Lavie Y, Tang X (1999) Localization and possible functions of phospholipase D isozymes. Biochim Biophys Acta 1439:245-263

Litosch I (2000) Regulation of phospholipase C-β_1 activity by phosphatidic acid. Biochem 39:7736-7743

Liu MY, Gutowski S, Sternweis PC (2001) The C-terminus of mammalian PLD is required for catalytic activity. J Biol Chem 276:5556-5562

Lopez I, Burns DJ, Lambeth JD (1995) Regulation of phospholipase D by protein kinase C in human neutrophils. Conventional isoforms of protein kinase C phosphorylate a phospholipase D-related component in the plasma membrane. J Biol Chem 270:19465-19472

Lopez I, Arnold RS, Lambeth JD (1998) Cloning and initial characterization of a human phospholipase D2 (hPLD2). J Biol Chem 273:12846-12852

Lu Z, Hornia A, Joseph T, Sukezane T, Frankel P, Zhong M, Bychenok S, Xu L, Feig LA, Foster DA (2000) Phospholipase D and RalA cooperate with the epidermal growth factor receptor to transform 3Y1 rat fibroblasts. Molec Cell Biol 20:462-467

Lucas L, del Peso P, Rodriguez P, Penalva V, Lacal JC (2000) Ras protein is involved in the physiological regulation of phospholipase D by platelet derived growth factor. Oncogene 19:431-437

Lukowski S, Lecomte M-C, Mira J-P, Marin P, Gautero H, Russo-Marie F, Geny B (1996) Inhibition of phospholipase D activity by fodrin. An active role for the cytoskeleton. J Biol Chem 271:24164-24171

Lukowski S, Mira J-P, Zachowski A, Geny B (1998) Fodrin inhibits phospholipases A2,C, and D by decreasing polyphosphoinositide cell content. Biochem Biophys Res Commun 248:278-284

Luo J-Q, Liu X, Hammond SH, Colley WC, Feig LA, Frohman MA, Morris AJ, Foster DA (1997) RalA interacts directly with the Arf-responsive PIP2-dependent phospholipase D1. Biochem Biophys Res Commun 235:854-859

Luo J-Q, Liu X, Frankel P, Rotunda T, Ramos M, Flom J, Jiang H, Feig LA, Morris AJ, Kahn RA, Foster DA (1998) Functional association between Arf and RalA in active phospholipase D complex. Proc Natl Acad Sci USA 95:3632-3637

MacDonald JIS, Sprecher H (1989) Distribution of arachidonic acid in choline- and ethanolamine-containing phosphoglycerides in subfractionated human neutrophils. J Biol Chem 264:17718-17726

MacNulty EE, Plevin R, Wakelam MJO (1990) Stimulation of the hydrolysis of phosphatidylinositol 4,5-bisphosphate and phosphatidylcholine by endothelin, a complete mitogen for Rat-1 fibroblasts. Biochem J 272:761-766

McPhail LC, Qualliotine-Mann D, Waite KA (1995) Cell-free activation of neutrophil NADPH oxidase by a phosphatidic acid-regulated protein kinase. Proc Natl Acad Sci USA 92:7931-7935

Mahadevappa VG, Holub BJ (1982) The molecular species composition of individual diacyl phospholipids in human platelets. Biochim Biophys Acta 713:73-79

Malcolm K, Elliott CM, Exton JH (1996) Evidence for Rho-mediated agonist stimulation of phospholipase D in Rat1 fibroblasts. Effects of *Clostridium Botulinum* C3 exoenzyme. J Biol Chem 271:13135-13139

Manifava M, Sugars J, Ktistakis NT (1999) Modification of catalytically active phospholipase D1 with fatty acid in vivo. J Biol Chem 274:1072-1077

Mao J, Yuan H, Xie W, Wu D (1998) Guanine nucleotide exchange factor GEF115 specifically mediates activation of Rho and serum response factor by the G protein α subunit G(13.Proc Natl Acad Sci USA 95:12973-12976

Marcil J, Harbour D, Naccache PH, Bourgoin S (1997) Human phospholipase D1 can be tyrosine-phosphorylated in HL-60 granulocytes. J Biol Chem 272:20660-20664

Martin TFJ, Hsieh K-P, Porter BW (1990) The sustained second phase of hormone-stimulated diacylglycerol accumulation does not activate protein kinase C in GH$_3$ cells. J Biol Chem 265:7623-7631

Martinson EA, Goldstein D, Brown JH (1989) Muscarinic receptor activation of phosphatidyl-choline hydrolysis. Relationship to phosphoinositide hydrolysis and diacylglycerol metabolism. J Biol Chem 264:14748-14754

Martinson E, Scheible S, Presek P (1994) Inhibition of phospholipase D of human platelets by protein tyrosine kinase inhibitors. Cell Mol Biol 40:627-634

Massenburg D, Han J-S, Liyanage M, Patton WA, Rhee SG, Moss J, Vaughan M (1994) Activation of rat brain phospholipase D by ADP-ribosylation factors 1, 5, and 6: separation of ADP-ribosylation factor-dependent and oleate-dependent enzymes. Proc Natl Acad Sci USA 91:11718-11722

Matozaki T, Williams JA (1989) Multiple sources of 1,2-diacylglycerol in isolated rat pancreatic acini stimulation by cholecystokinin. Involvement of phosphatidylinositol bisphosphate and phosphatidylcholine hydrolysis. J Biol Chem 264:14729-14734

Meacci E, Vasta V, Moorman JP, Bobak DA, Bruni P, Moss J, Vaughan M (1999) Effect of Rho and ADP-ribosylation factor GTPases on phospholipase D activity in intact human adenocarcinoma A549 cells. J Biol Chem 274:18605-18612

Meier KE, Gibbs TC, Knoepp SM, Ella KM (1999) Expression of phospholipase D isoforms in mammalian cells. Biochim Biophys Acta 1439:199-213

Michaely PA, Mineo C, Ying Y-s, Anderson RGW (1999) Polarized distribution of endogenous Rac1 and RhoA at the cell surface. J Biol Chem 274:21430-21436

Min DS, Exton JH (1998) Phospholipase D is associated in a phorbol ester-dependent manner with protein kinase C-α and with a 220-kDa protein which is phosphorylated on serine and threonine. Biochem Biophys Res Commun 248:533-537

Min DS, Kim E-G, Exton JH (1998a) Involvement of tyrosine phosphorylation and protein kinase C in the activation of phospholipase D by H$_2$O$_2$ in Swiss 3T3 fibroblasts. J Biol Chem 273:29986-29994

Min DS, Park S-K, Exton JH (1998b) Characterization of a rat brain phospholipase D isozyme. J Biol Chem 273:7044-7051

Min DS, Cho NJ, Yoon SH, Lee YH, Hahn S-J, Lee K-H, Kim M-S, Jo Y-H (2000) Phospholipase C, protein kinase C, Ca^{2+}/calmodulin-dependent protein kinase II, and tyrosine phosphorylation are involved in carbachol-induced phospholipase D activation in Chinese hamster ovary cells expressing muscarinic acetylcholine receptor of *Caenorhabditis elegans*. J Neurochem 75:274-281

Mitchell R, McCulloch D, Lutz E, Johnson M, MacKenzie C, Fennell M, Fink G, Zhou W, Sealfon SC (1998) Rhodopsin-family receptors associate with small G proteins to activate phospholipase D. Nature 392:411-414

Morash SC, Roose SD, Byers DM, Ridgway ND, Cook HW (1998) Overexpression of myristoylated alanine-rich C-kinase substrate enhances activation of phospholipase D by protein kinase C in SK-N-MC human neuroblastoma cells. Biochem J 332:321-327

Morash SC, Byers DM, Cook HW (2000) Activation of phospholipase D by PKC and GTPγS in human neuroblastoma cells overexpressing MARCKS. Biochim Biophys Acta 1487:177-189

Morgan CP, Sengelov H, WhatmoreJ, Borregaard N, Cockcroft S (1997) ADP-ribosylation-factor-regulated phospholipase D activity localizes to secretory vesicles and mobilizes to the plasma membrane following N-formylmethionyl-leucyl-phenylalanine stimulation of human neutrophils. Biochem J 325:581-585

Moritz A, De Graan PNE, Gispen WH, Wirtz KWA (1992) Phosphatidic acid is a specific activator of phosphatidylinositol-4-phosphate kinase. J Biol Chem 267:7207-7210

Moss J, Vaughan M (1998) Molecules in the ARF orbit. J Biol Chem 273:21431-21434

Munnik T (2001) Phosphatidic acid: an emerging plant lipid second messenger. Trends Plant Sci 6:227-233

Murthy KS, Zhou H, GriderJR, Makhlouf GM (2001) Sequential activation of heterotrimeric and monomeric G proteins mediates PLD activity in smooth muscle. Am. J. Physiol 280:G381-G388

Musial A, Mandal A, Coroneos E, Kester M (1995) Interleukin-1 and endothelin stimulate distinct species of diglycerides that differentially regulate protein kinase C in mesangial cells. J Biol Chem 270:21632-21638

Muthalif MM, Parmentier J-H, Benter IF, Karzoun N, Ahmed A, Khandekar Z, Adl MZ, Bourgoin S, Malik KU (2000) Ras/mitogen-activated protein kinase mediates norepinephrine-induced phospholipase D activation in rabbit aortic smooth muscle cells by a phosphorylation-dependent mechanism. J Pharmacol Exper Therap 293:268-274

Nakamura T, Abe A, Balazovich KJ, Wu D, Suchard SJ, Boxer LA, Shayman JA (1994) Ceramide regulates oxidant release in adherent human neutrophils. J Biol Chem 269:18384-18389

Natarajan V, Taher MM, Roehm B, Parinandi NL, Schmid HHO, Kiss Z, Garcia JGN (1993) Activation of endothelial cell phospholipase D by hydrogen peroxide and fatty acid hydroperoxide. J Biol Chem 268:930-937

Natarajan V, Scribner WM, Vepa S (1996) Regulation of phospholipase D by tyrosine kinases. Chem. Physics Lipids 80:103-116

Nixon JS (1997) The biology of protein kinase C inhibitors. Eds.: Parker PJ & Dekker LV, R.G. Landes Company (Austin), pp 205-236

Nobes C, Hall A (1994) Regulation and function of the Rho subfamily of small GTPases. Curr Opin Genet Develop 4:77-81

Offermanns S, Bombien E, Schultz G (1993) Thrombin Ca^{2+}-dependently stimulates protein tyrosine phosphorylation in BC_3H1 muscle cells. Biochem J 290:27-32

Ohguchi K, Banno Y, Nakashima S, Nozawa Y (1996) Regulation of membrane-bound phospholipase D by protein kinase C in HL60 cells. J Biol Chem 271:4366-4372

Ohguchi K, Kasai T, Nozawa Y (1997) Tyrosine phosphorylation of 100-115kDa proteins by phosphatidic acid generated via phospholipase D activation inHL60 granulocytes. Biochim Biophys Acta 1346:301-304

Ohtsuka T, Ozawa M, Okamura N, Ishibashi S (1989) Stimulatory effects of a short chain phosphtidate on superoxide anion production in guinea pig polymorphonuclear leukocytes. J Biochem 106:259-263

Oishi K, Takahashi M, Mukai H, Banno Y, Nakashima S, Kanaho Y, Nozawa Y, Ono Y (2001) PKN regulates phospholipase D1 through the direct interaction. J Biol Chem In press

Ojio K, Banno Y, Nakashima S, Kato N, Watanabe K, Lyerly DM, Miyata H, Nozawa Y (1996) Effect of *Clostridium difficile* Toxin B on IgE receptor-mediated signal transduction in rat basophilic leukemia cells: Inhibition of phospholipase D activation. Biochem Biophys Res Commun 224:591-596

Olivier AR, Parker PJ (1992) Identification of multiple PKC isoforms in Swiss 3T3 cells: Differential down-regulation by phorbol ester. J Cell Physiol 152:240-244

Pachter, JA, Pai J-K, Mayer-Ezell R, Petrin JM, Dobek E, Bishop WR (1992) Differential regulation of phosphoinositide and phosphatidylcholine hydrolysis by protein kinase C-β1 overexpression. Effects on stimulation by α-thrombin, guanosine 5'-O-(thiotriphosphate), and calcium. J Biol Chem 267:9826-9830

Pai J-K, Siegel MI, Egan RW, Billah MM (1988) Phospholipase D catalyzes phospholipid metabolism in chemotactic peptide-stimulated HL-60 granulocytes. J Biol Chem 263:12472-12477

Pai J-K, Dobek EA, Bishop WR (1991) Endothelin-1 activates phospholipase D and thymidine incorporation in fibroblasts overexpressing protein kinase C(1. Cell Regul 2:897-903

Palicz A, Foubert TR,Jesaitis AJ, Marodi L, McPhail LC (2001) Phosphatidic acid and diacylglycerol directly activate NADPH oxidase by interacting with enzyme components. J Biol Chem 276:3090-3097

Pappan K, Wang X (1999) Molecular and biochemical properties and physiological roles of plant phospholipase D. Biochim Biophys Acta 1439:151-166

Pappan K, Zheng S, Wang X (1997) Identification and characterization of a novel plant phospholipase D that requires polyphosphoinositides and submicromolar calcium for activity in Arabidopsis. J Biol Chem 272:7048-7054

Park JB, Kim JH, Kim Y, Ha SH, Kim JH, Yoo J-S, Du G, Frohman MA, Suh P-G, Ryu SH (2000) Cardiac phospholipase D2 localizes to sarcolemmal membranes and is inhibited by α-actinin in an ADP-ribosylation factor-reversible manner. J Biol Chem 275:21295-21301

Park S-K, Provost JJ, Bae CD, Ho W-T, Exton JH (1997) Cloning and characterization of phospholipase D from rat brain. J Biol Chem 272:29263-29271

Park S-K, Min DS, Exton JH (1998) Definition of the protein kinase C interaction site of phospholipase D. Biochem Biophys Res Commun 244:364-367

Parmentier J-H, Muthalif MM, Saeed AE, Malik KU (2001) Phospholipase D activation by norepinephrine is mediated by 12(S)-, 15(S)-, and 20-hydroxyeicosatetraeonoic acids generated by stimulation of cytosolic phospholipase A$_2$. Tyrosine phosphorylation of phospholipase D$_2$ in response to norepinephrine. J Biol Chem 276:15704-15711

Patton GM, Fasulo JM, Robins SJ (1982) Separation of phospholipids and individual molecular species of phospholipids by high-performance liquid chromatography. J Lipid Res 23:190-196

Pertile P, Liscovitch M, Chalifa V, Cantley LC (1995) Phosphatidylinositol 4,5-bisphosphate synthesis is required for activation of phospholipase D in U937cells. J Biol Chem 270:5130-5135

Pessin MS, Raben DM (1989) Molecular species analysis of 1,2-diglycerides stimulated by α-thrombin in cultured fibroblasts. J Biol Chem 264:8729-8738

Pessin MS, Baldassare JJ, Raben DM (1990) Molecular species analysis of mitogen-stimulated 1,2-diglycerides in fibroblasts. Comparison of α-thrombin, epidermal growth factor, and platelet-derived growth factor. J Biol Chem 265:7959-7966

Pete MJ, Ross AH, Exton JH (1994) Purification and properties of phospholipase A_1 from bovine brain. J Biol Chem 269:19494-19500

Pettit TR, Martin A, HortonT, Liossis C, Lord, JM, Wakelam MJO (1997) Diacylglycerol and phosphatidate generated by phospholipases C and D, respectively, have distinct fatty acid compositions and functions. Phospholipase D-derived diacylglycerol does not activate protein kinase C in porcine aortic endothelial cells. J Biol Chem 272:17354-17359

Pfeilschifter J, Huwiler A (1993) A role for protein kinase C-ε in angiotensin II stimulation of phospholipase D in rat renal mesangial cells. FEBS Lett 331:267-271

Plevin R, Cook SJ, Palmer S, Wakelam MJO (1991) Multiple sources of sn-1,2-diacylglycerol in platelet-derived-growth-factor-stimulated Swiss 3T3 fibroblasts. Biochem J 279:559-565

Plonk SG, Park S-K, Exton JH (1998) The α-subunit of the heterotrimeric G protein G_{13} activates a phospholipase D isozyme by a pathway requiring Rho family GTPases. J Biol Chem 273:4823-4826

Ponting CP, Kerr ID (1996) A novel family of phospholipase D homologues that includes phospholipid synthases and putative endonucleases: identification of duplicated repeats and potential active-site residues. Protein Sci 5:914-922

Popoff MR, Chaves-Olarate E, Lemichez E, von Eichel-Streiber C, Thelestam M, Chardin P, Cussac D, Antonny B, Chavrier P, Flatau G, Giry M, de Gunzburg J, Boquet P (1996) Ras, Rap, and Rac small GTP-binding proteins are targets for Clostridium sordellii lethal toxin glucosylation. J Biol Chem 271:10217-10224

Prenzel N, Zwick E, Daub H, Leserer M, Abraham R, Wallasch C, Ullrich A (1999) EGF Receptor transactivation by G-protein-coupled receptors requires metallo-proteinase cleavage of proHB-EGF. Nature 402:884-888

Price BD, Morris JDH, Hall A (1989a) Stimulation of phosphatidylcholine breakdown and diacylglycerol production by growth factors in Swiss 3T3cells. Biochem J 264:509-515

Price BD, Morris JDH, Marshall CJ, Hall A (1989b) Stimulation of phosphatidylcholine hydrolysis, diacylglycerol release, and arachidonic acid production by oncogenic Ras is a consequence of protein kinase C activation. J Biol Chem 264:16638-16643

Provost JJ, Fudge J, Israelit S, Siddiqi AR, Exton JH (1996) Tissue-specific distribution and subcellular distribution of phospholipase D in rat: evidence for distinct RhoA- and ADP -ribosylation factor (ARF)-regulated isoenzymes. Biochem J 319:285-291

Qin W, Pappan K, Wang X (1997) Molecular heterogeneity of phospholipase D (PLD). Cloning of PLDγ and regulation of plant PLDγ, -β, and -α by polyphosphoinositides. J Biol Chem 272:28267-28273

Qualliotine-Mann D, Agwu DE, Ellenburg MD, McCall CE, McPhail LC (1993) Phos-
    phatidic acid and diacylglycerol synergize in a cell-free system for activation of
    NADPH oxidase from human neutrophils. J Biol Chem 268:23843-23849
Quilliam LA, Der CJ, Brown JH (1990) GTP-binding protein-stimulated phospholi-
    pase D and phospholipase D activities in ras-transformed NIH 3T3 fibroblasts.
    Second Messengers and Phosphoproteins 13:59-67
Radhakrishna H, Donaldson JG (1997) ADP-ribosylation factor 6 regulates a novel
    plasma membrane recycling pathway. J Cell Biol 139:49-61
Regier DS, Waite KA, WallilnR, McPhail LC (1999) A phosphatidic acid-activated
    protein Kinase and conventional protein kinase C isoforms phosphorylate
    p22^{phox}, an NADPH oxidase component. J Biol Chem 274:36601-36608
Regier DS, Green DG, Sergeant S, Jesaitis AJ, McPhail LC (2000) Phosphorylation of
    p22^{phox} is mediated by phospholipase D-dependent and -independent mecha-
    nisms. Correlation of NADPH oxidase activity and p22^{phox} phosphorylation. J
    Biol Chem 275:28406-28412
Ren X-D, Bokoch GM, Traynor-Kaplan A, Jenins GH, Anderson RA, Schwartz MA
    (1996) Physical association of the small GTPase Rho with a 68-kDa phosphatidy-
    linositol 4-phosphate 5-kinase in Swiss 3T3 cells. Mol Biol Cell 7:435-442
Rizzo MA, Shome K, Watkins SC, Romero G (2000) The recruitment of Raf-1 to
    membranes is mediated by direct interaction with phosphatidic acid and is inde-
    pendent of association with Ras. J Biol Chem 275:23911-23918
Rossi F, Grzeskowiak M, Della-Bianca V, Calzetti F, Gandini G (1990) Phosphatidic
    acid and not diacylglycerol generated by phospholipase D is functionally linked
    to the activation of the NADPH oxidase by fMLP in human neutrophils. Biochem
    Biophys Res Commun 168:320-327
Rudge SA, Morris AJ, Engebrecht JA (1998) Relocalization of phospholipase D activ-
    ity mediates membrane formation during meiosis. J Cell Biol 140:81-90
Rudge SA, Engebrecht J (1999) Regulation and function of PLDs in yeast. Biochim
    Biophys Acta 1439:167-174
Rudolph AE, Stuckey JA, Zhao Y, Matthews HR, Patton WA, Moss J, Dixon JE (1999)
    Expression, characterization, and mutagenesis of the Yersinia pestis murine
    toxin, a phospholipase D superfamily member. J Biol Chem 274:11824-11831
Rümenapp U, Geiszt M, Wahn F, Schmidt M, Jakobs KH (1995) Evidence for ADP-
    ribosylation-factor-mediated activation of phospholipase D by m3 muscarinic
    acetylcholine receptor. Eur J Biochem 234:240-244
Rümenapp U, Schmidt M, Wahn F, Tapp E, Grannass A, Jakobs KH (1997) Charac-
    teristics of protein-kinase-C- and ADP-ribosylation-factor-stimulated phospho-
    ipase D activities in human embryonic kidney cells. Eur J Biochem 248:407-414
Rümenapp U, Asmus M, Schablowski H, Woznicki M, Han L, Jakobs KH, Fahimi-
    Vahid M, Michalek C, Wieland T, Schmidt M (2001) The M3 muscarinic acetyl-
    choline receptor expressed in HEK-293 cells signals to phospholipase D via G_{12}
    but not G_q-type G proteins. regulators of G proteins as tools to dissect pertussis
    toxin-resistant G proteins in receptor-effector coupling. J Biol Chem 276:2474-
    2479
Schieven GL, Kirihara JM, Burg DL, Geahlen RL Ledbetter JA (1993) p72^{syk} tyrosine
    kinase is activated by oxidizing conditions that induce lymphocyte tyrosine
    phosphorylation and Ca^{2+} signals. J Biol Chem 268:16688-16692
Schmidt M, Rümenapp U, Bienek C, Keller J, von Eichel-Streiber C, Jakobs KH
    (1996a) Inhibition of receptor signaling to phospholipase D by Clostridium dif-
    ficile Toxin B. Role of Rho proteins. J Biol Chem 271:2422-2426

Schmidt M, Rümenapp U, Nehls C, Ott S, Keller J, von Eichel-Streiber C, Jakobs KH (1996b) Restoration of *Clostridium difficile* toxin-B-inhibited phospholipase D by phosphatidylinositol 4,5-bisphosphate. Eur J Biochem 240:707-712

Schmidt M, Voss M, Thiel M, Baur B, Grannas A, Tapp E, Cool RH, de Gunzburg J, von Eichel-Streiber C, Jakobs KH (1998) Specific inhibition of phorbol ester-stimulated phospholipase D by *Clostridium sordellii* lethal toxin and *Clostridiuim difficile* toxin B-1470 in HEK-293 cells. J Biol Chem 273:7413-7422

Schmidt M, Voß M, Oude Weernink PA, Wetzel J, Amano M, Kaibuchi K, Jakobs KH (1999) A role for Rho-kinase in Rho-controlled phospholipase D stimulation by the m3 muscarinic acetylcholine receptor. J Biol Chem 274:14648-14654

Schmidt M, Hüwe SM, Fasselt B, Homann D, Rümenapp U, Sandmann J, Jakobs KH (1994) Mechanisms of phospholipase D stimulation by m3 muscarinic acetylcholine receptors. Evidence for involvement of tyrosine phosphorylation. Eur J Biochem 225:667-675

Schürmann A, Schmidt M, Asmus M, Bayer S, Fliegert F, Koling S, Massmann S, Schilf C, Subauste MC, Voss M, Jakobs KH, Joost H-G (1999) The ADP-ribosylation factor (ARF)-related GTPase ARF-related protein binds to the ARF-specific guanine nucleotide exchange factor cytohesin and inhibits the ARF-dependent activation of phospholipase D. J Biol Chem 274:9744-9751

Sciorra VA, Morris AJ (1999) Sequential actions of phospholipase D and phosphatidic acid phosphohydrolase 2b generate diglyceride in mammalian cells. Molec Biol Cell 10:3863-3876

Sciorra VA, Hammond SM, Morris AJ (2001) Potent direct inhibition of mammalian phospholipase D isoenzymes by calphostin-c. Biochem 40:2640-2646

Sciorra VA, Rudge SA, Prestwich GD, Frohman MA, Engebrecht JA, Morris AJ (1999) Identification of a phosphoinositide binding motif that mediates activation of mammalian and yeast phospholipase D isoenzymes EMBO J 20:5911-5921

Sebaldt RJ, Adams DO, Uhing RJ (1992) Quantification of contributions of phospholipid precursors to diradylglycerols in stimulated mononuclear phagocytes. Biochem J 284:367-375

Sechi AS, Wehland J (2000) The actin cytoskeleton and plasma membrane connection: PtdIns $(4,5)P_2$ influences cytoskeletal protein activity at the plasma membrane. J Cell Sci 113:3685-3695

Senogles SE (2000) The D2 dopamine receptor stimulates phospholipase D activity: A novel signaling pathway for dopamine. Molec Pharmacol 58:455-462

Sergeant S, Waite KA, Heravi J, McPhail LC (2001) Phosphatidic acid regulates tyrosine phosphorylating activity in human neutrophils. Enhancement of Fgr activity. J Biol Chem 276:4737-4746

Shen Y, Xu, L, Foster DA (2001) Role for phospholipase D in receptor-mediated endocytosis. Molec Cell Biol 21:595-602

Shome K, Vasudevan C, Romero G (1997) ARF proteins mediate insulin-dependent activation of phospholipase D. Curr Biol 7:387-396

Shome K, Nie Y, Romero G (1998) ADP-ribosylation factor proteins mediate agonist-induced activation of phospholipase D. J Biol Chem 273:30836-30841

Shome K, Rizzo MA, Vasudevan C, Andresen B, Romero G (1999) The activation of phospholipase D by endoethelin-1, angiotensin II, and platelet-derived growth factor in vascular smooth muscle A10 cells is mediated by small G proteins of the ADP-ribosylation factor family. Endocrinol 141:2200-2208

Siddhanta A, Shields D (1998) Secretory vesicle budding from the trans-Golgi network is mediated by phosphatidic acid levels. J Biol Chem 273:17995-17998

Siddhanta A, Backer JM, Shields D (2000) Inhibition of phosphatidic acid synthesis alters the structure of the Golgi apparatus and inhibits secretion in endocrine cells. J Biol Chem 275:12023-12031

Siddiqi AR, Srajer GE, Leslie CC (2000) Regulation of human PLD1 and PLD2 by calcium and protein kinase C. Biochim Biophys Acta 1497:103-114

Siddiqui RA, Yang Y-C (1995) Interleukin-11 induces phosphatidic acid formation and activates MAP kinase in mouse 3T3-L1 cells. Cell Signal 7:247-259

Singer WD, Brown HA, Bokoch GM, Sternweis PC (1995) Resolved phospholipase D activity is modulated by cytosolic factors other than Arf. J Biol Chem 270:14944-14950

Singer WD, Brown HA, Jiang X, Sternweis PC (1996) Regulation of phospholipase D by protein kinase C is synergistic with ADP-ribosylation factor and independent of protein kinase activity. J Biol Chem 271:4504-4510

Sinnett-Smith J, Zachary I, Valverde AM, Rozengurt E (1993) Bombesin stimulation of p125 focal adhesion kinase tyrosine phosphorylation. Role of protein kinase C, Ca^{2+} mobilization and the actin cytoskeleton. J Biol Chem 268:14261-14268

Slaaby R, Jensen T, Hansen HS, Frohman MA, Seedorf K (1998) PLD2 complexes with the EGF receptor and undergoes tyrosine phosphorylation at a single site upon agonist stimulation J Biol Chem 273:33722-33727

Slaaby R, Du G, Altshuller YM, Frohman MA, Seedorf K (2000) Insulin-induced phospholipase D1 and phospholipase D2 activity in human embryonic kidney-293 cells mediated by the phospholipase Cγ and protein kinase Cα signalling cascade. Biochem J 351:613-619

Slater SJ, Seiz JL, Stagliano BA, Cook AC, Milano SK, Ho C, Stubbs CD (2001) Low- and high-affinity phorbol esterand diglyceride interactions with protein kinaseC: 1-O-Alkyl-2 -acyl-sn-glycerol enhances phorbol ester- and diacylglycerol-induced activity but alone does not induce activity. Biochem 40:6085-6092

Slivka SR, Meier KE, Insel PA (1988) (1-Adrenergic receptors promote phosphatidylcholine hydrolysis in MDCK-D1 cells. A mechanism for rapid activation of protein kinase C. J Biol Chem 263:12242-12246

Smith M, Jungalwala FB (1981) Reversed-phase high performance liquid chromatography of phosphatidylcholine: a simple method for determining relative hydrophobic interaction of various molecular species. J Lipid Res 22:697-704

Song J, Pfeffer LM, Foster DA (1991) v-Src increases diacylglycerol levels via a type D phospholipase-mediated hydrolysis of phosphatidylcholine. Mol Cell Biol 11:4903-4908

Spiegel S, Foster D, Kolesnick R (1996) Signal transduction through lipid second messengers. Curr Opin Cell Biol 8:159-167

Stamnes MA, Rothman JE (1993) The binding of AP-1 clathrin adaptor particles to Golgi membranes requires ADP ribosylation factor, a small GTP-binding protein. Cell 73:999-1005

Stamnes M, Schiavo G, Stenbeck G, Sollner TH, Rothman JE (1998) ADP-ribosylation factor and phosphatidic acid levels in Golgi membranes during budding of coatomer-coated vesicles. Proc Natl Acad Sci USA 95:13676-13680

Stanacev NZ, Stuhne-Sekalec L (1970) On the mechanisms of enzymatic phosphatidylation. Biosynthesis of cardiolipin catalyzed by phospholipase D. Biochim Biophys Acta 210:350-352

Stuckey JA, Dixon JE (1999) Crystal structure of a phospholipase D family member. Nature Struct Biol 6:278-284

Stutchfield J, Cockcroft S (1993) Correlation between secretion and phospholipase D activation in differentiated HL60cells. Biochem J 293:649-655

Sugars JM, Cellek S, Manifava M, Coadwell J, Ktistakis NT (1999) Fatty acylation of phospholipase D1 on cysteine residues 240 and 241 determines localization on intracellular membranes. J Biol Chem 274:30023-30027

Sung T-C, Roper RL, Zhang Y, Rudge SA, Temel R, Hammond SM, Morris AJ, Moss B, Engebrecht JA, Frohman MA (1997) Mutagenesis of phospholipase D defines a superfamily including a *trans*-Golgi viral protein required for poxvirus pathogenicity. EMBO J 16:4519-4539

Sung T-C, Altshuller YM, Morris AJ, Frohman MA (1999a) Molecular analysis of mammalian phospholipase D2. J Biol Chem 274:494-502

Sung T-C, Zhang Y, Morris AJ, Frohman MA (1999b) Structural analysis of human phospholipase D1. J Biol Chem 274:3659-3666

Suzuki A, Kozawa O, Shinoda J, Watanabe Y, Saito H, Oiso Y (1996a) Thrombin induces proliferation of osteoblast-like cells through phosphatidylcholine hydrolysis. J Cell Physiol 168:209-216

Suzuki A, Shinoda J, Oiso Y, Kozawa O (1996b) Tyrosine kinase is involved in angiotensin II-stimulated phospholipase D activation in aortic smooth muscle cells: Function of Ca^{2+} influx. Atherosclerosis 121:119-127

Takamura H, Narita H, Park HJ, Tanaka K-I, Matsuura T, Kito M (1987) Differential hydrolysis of phospholipid molecular species during activation of human platelets with thrombin and collagen. J Biol Chem 262:2262-2269

Tang H, Zhao ZJ,, Landon EJ, Inagami T (2000) Regulation of calcium-sensitive tyrosine kinase Pyk2 by angiotensin II in endothelial cells. Roles of Yes tyrosine kinase and tyrosine phosphatase SHP-2. J Biol Chem 275:8389-8396

Tapia JA, Ferris HA, Jensen RT, Garcia LJ (1999) Cholecystokinin activates PYK2/CAKβ by a phospholipase C-dependent mechanism and its association with the mitogen-activated protein kinase signaling pathway in pancreatic acinar cells. J Biol Chem 274:31261-31271

Thorsen VAT, Bjorndal B, Nolan G, Fukami MH, Bruland O, Lillehaug JR, Holmsen H (2000) Expression of a peptide binding to receptor for activated C-kinase (RACK1) inhibits phorbol myristoyl acetate-stimulated phospholipase D activity in C3H/10T1/2 cells: dissociation of phospholipase D-mediated phosphatidylcholine breakdown from its synthesis. Biochim Biophys Acta 1487:163-176

Tomic S, Greiser U, Lammers R, Kharitonenkov A, Imyanitov E, Ullrich A, Böhmer F-D (1995) Association of SH2 domain protein tyrosine phosphatases with the epidermal growth factor receptor in human tumor cells. Phosphatidic acid activates receptor dephosphorylation by PTP1C. J Biol Chem 270:21277-21284

Tool ATJ, Blom M, Roos D, Verhoeven AJ (1999) Phospholipase D-derived phosphatidic acid is involved in the activation of the CD11b /CD18 integrin in human eosinophils. Biochem J 340:95-101

Traub LM, Ostrom JA, Kornfeld S (1993) Biochemical dissection of AP-1 recruitment onto Golgi membranes. J Cell Biol 135:1801-1814

Tronchère H, Planat V, Record M, Tercé F, Ribbes G, Chap H (1995) Phosphatidylcholine turnover in activated human neutrophils. Agonist-induced cytidyltransferase translocation is subsequent to phospholipase D activation. J Biol Chem 270:13138-13146

Tsai M-H, Yu C-L, Wie F-S, Staacey DW (1989) The effect of GTPase activating protein upon Ras is inhibited by mitogenically responsive lipids. Science Wash. DC 243:522-526

Tsai S-C, Adamik R, Hong J-X, Moss J, Vaughan M, Kanoh H, Exton JH (1998) Effects of arfaptin 1 on guanine nucleotide-dependent activation of phospholipase D and cholera toxin by ADP-ribosylation factor. J Biol Chem 273:20697-20701

Tüscher O, Lorra C, Bouma B, Wirtz KWAA, Huttner WB (1997) Cooperativity of phosphatidyliositol transfer protein and phospholipase D in secretory vesicle formation from the TGN – phosphoinositides as a common denominator? FEBS Lett 419:271-275

Uings IJ, Thompson NT, Randall RW, Spacey GD, Bonser RW, Hudson, Garland LG (1992) Tyrosine phosphorylation is involved in receptor coupling to phospholipase D but not phospholipase C in the human neutrophil. Biochem J 281:597-600

van Blitterswijk WJ, Hilkmann H, de Widt J, van der Bend RL (1991a) Phospholipid metabolism in bradykinin-stimulated human fibroblasts. I. Biphasic formation of diacylglycerol from phosphatidylinositol and phosphatidylcholine, controlled by protein kinase C. J Biol Chem 266:10337-19343

van Blitterswijk WJ, Hilmann H, de Widt J, van der Bend, RL (1991b) Phospholipid metabolism in bradykinin-stimulated human fibroblasts. II. Phosphatidylcholine breakdown by phospholipases C and D; involvement of protein kinase C. J Biol Chem 266:10344-10350

van Dijk MCM, Postma F, Hilkmann H, Jalink K, van Blitterswijk WJ, Moolenaar WH (1998) Exogenous phospholipase D generates lysophosphatidic acid and activates Ras, Rho and Ca^{2+} signaling pathways. Curr Biol 8:386-392

Venable ME, Blobe, GC, Obeid LM (1994) Identification of a defect in the phospholipase D/diacylglycerol pathway in cellular senescence. J Biol Chem 269:26040-26044

Vieira AV, Lamaze C, Schmid SL (1996) Control of EGF receptor signaling by clathrin-mediated endocytosis. Science 274:2086-2089

Vitale N, Caumont A-S, Chasserot-Golaz S, Du G, Wu S, Sciorra VA, Morris AJ, Frohman MA, Bader M-F (2001) Phospholipase D1: a key factor for the exocytotic machinery in neuroendocrine cells. EMBO J 20:2424-2434

von Eichel-Streiber C, Boquet P, Sauerborn M, Thelestam M (1996) Large clostridial cytotoxinsa family of glycosyltransferases modifying small GTP-binding proteins. Trends Microbiol4:375-382

Voss M, Oude Oude Weernink PA, Haupenthal S, Moller U, Cool RH, Bauer B, Camonis JH, Jakobs KH,Schmidt M (2000) Phospholipase D stimulation by receptor tyrosine kinases mediated by protein kinase C and a Ras/Ral signaling cascade. J Biol Chem 274:34691-34698

Waite KA, Wallin R, Qualliotine-Mann D, McPhail LC (1997) Phosphatidic acid-mediated phosphorylation of the NADPH oxidase component p47-*phox*. Evidence that phosphatidic acid may activate a novel protein kinase. J Biol Chem 272:15569-15578

Waite M (1999) The PLD superfamily: insights into catalysis. Biochim Biophys Acta 1439:187-197

Wakelam MJO (1998) Diacylglycerol – when is it an intracellular messenger? Biochim Biophys Acta 1436:117-126

Walker SJ, Wu W-J, Cerione RA, Brown HA (2000) Activation of phospholipase D1 by Cdc42 requires the Rho insert region. J Biol Chem 275:15665-15668

Wang S, Banno Y, Nakashima S, Nozawa, Y (2001) Enzymatic characterization of phospholipase D of protozoan *Tetrahymena* Cells. J Eukaryot Microbiol 48:194-201

Ward DT, Ohanian J, Heagerty AM, Ohanian V (1995) Phospholipase D-induced phosphatidate production in intact small arteries during noradrenaline stimulation: involvement of both G-protein and tyrosine-phosphorylation-linked pathways. Biochem J 307:451-456.

Watanabe H, Kanaho Y (2000) Inhibition of phosphatidylinositol 4,5-bisphosphate-stimulated phospholipase D2 activity by Ser/Thr phosphorylation. Biochim Biophys Acta 1495:121-124

Way G, O'Luanaigh N, Cockcroft S (2000) Activation of exocytosis by cross-linking of the IgE receptor is dependent on ADP-ribosylation factor1-regulated phospholipase D in RBL-2H3 mast cells: evidence that the mechanism of activation is via regulation of phosphatidylinositol 4,5-bisphosphate synthesis. Biochem J 346:63-70

Oude Oude Weernink PA, Schulte P, Guo Y,Wetzel J, Amano M, Kaibuchi K,Haverland S, VoB M, Schmidt M, Mayr GW, Jakobs KH (2000) Stimulation of phosphatidylinositol-4-phosphate 5-kinase by Rho-kinase. J Biol Chem 275:10168-10174

West MA, Bright NA, Robinson MS (1997) The role of ADP-ribosylation factor and phospholipase D in adaptor recruitment. J Cell Bioll 138:1239-1254

Whatmore J, Morgan CP, Cunningham E, Collison KS, Willison KR, Cockcroft S (1996) ADP-ribosylation factor 1-regulated phospholipase D activity is localized at the plasma membrane and intracellular organelles in HL60 cell. Biochem J 320:785-794

Wilkie N, Morton C, Ng LL, Boarder MR (1996) Stimulated mitogen-activated protein kinase is necessary but not sufficient for the mitogenic response to angiotensin II. A role for phospholipase D. J Biol Chem 271:32447-32453

Wilkes LC, Patel V, Purkiss JR, Boarder MR (1993) Endothelin-1 stimulated phospholipase D in A10 vascular smooth muscle derived cells is dependent on tyrosine kinase. FEBS Lett 322:147-150

Williger B-T, Provost JJ, Ho W-T, Milstine J, Exton JH (1999) Arfaptin 1 forms a complex with ADP-ribosylation factor and inhibits phospholipase D. FEBS Lett 454:85-89

Williger B-T, HoW-T, Exton JH (1999) PhospholipaseD mediates matrix metaloproteinase-9 secretion in phorbol ester-stimulated human fibrosarcoma cells. J Biol Chem 274:735-738

Williger B-T, Ostermann J, Exton JH (1999) Arfaptin 1, an ARF-binding protein, inhibits phospholipase D and endoplasmic reticulum/Golgi protein transport. FEBS Lett 443:197-200

Wright TM, Rangan LA, Shin HS, Raben DM (1988) Kinetic analysis of 1,2-diacylglycerol mass levels in cultured fibroblasts. Comparison of stimulation by α-thrombin and epidermal growth factor. J Biol Chem 263:9374-9380

Xie Z, Ho W-T, Exton JH (1998) Association of N- and C-terminal domains of phospholipase D is required for catalytic activity. J Biol Chem 273:34679-34682

Xie Z, Ho W-T, Exton JH (2000a) Association of the N-and C-terminal domains of phospholipase D. Contribution of the conserved HKD motifs to the interaction and the requirement of the association for Ser/Thr phosphorylation of the enzyme. J Biol Chem 275:24962-24969

Xie Z, Ho W-T, Exton JH (2000b) Conserved amino acids at the C-terminus of rat phospholipase D1 are essential for enzymatic activity. Eur J Biochem 267:7138-7146

Xie Z, Ho W-T, Exton JH (2001) Requirements and effects of palmitoylation of rat PLD1. J Biol Chem 275:24962-24969

Yamazaki M, Zhang Y, Watanabe H, Yokozeki T, Ohno S, Kaibuchi K, Shibata H, Mukai H, Ono Y, Frohman MA, Kanaho Y (1999) Interaction of the small G protein RhoA with the C terminus of human phospholipase D1. J Biol Chem 274:6035-6038

Yang SF, Freer S, Benson AA (1967) Transphosphatidylation by phospholipase D. J Biol Chem 242:477-484

Yeo E-J, Kazlauskas A, Exton JH (1994) Activation of phospholipase C-γ is necessary for stimulation of phospholipase D by platelet-derived growth factor. J Biol Chem 269:27823-7826

Yu H, Li X, Marchetto GS, Dy R, Hunter D, Calvo B, Dawson TL, Wilm M, Anderegg RJ, Graves LM, Earp HS (1996) Activation of a novel calcium-dependent protein-tyrosine kinase. Correlation with c-Jun N-terminal kinase but not mitogen-activated protein kinase activation. J Biol Chem 271:29993-29996

Yuli I, Tomonaga A, Snyderman R (1982) Chemoattractant receptor functions in human polyorphonuclear leukocytes are divergently altered by membrane fluidizers. Proc Natl Acad Sci USA 79:5906-5910

Zhang G-F, Patton WA, Lee F-JL, Liyanage M, Han J-S, Rhee SG, Moss J, Vaughan M (1995) Different ARF domains are required for the activation of cholera toxin and phospholipase D. J Biol Chem 270:21-24

Zhang H, Desai NN, Murphey JM, Spiegel S (1990) Increases in phosphatidic acid levels accompany sphingosine-stimulated proliferation of quiescent Swiss 3T3 cells. J Biol Chem 265:21309-21316

Zhang Y, Altshuller YM, Hammond SM, Morris AJ, Frohman MA (1999) Loss of receptor regulation by a phospholipase D1 mutant unresponsive to protein kinase C. EMBO J18:6339-6348

Zhao Z, Shen S-H, Fischer EH (1993) Stimulation by phospholipids of a protein-tyrosine-phosphatase containing two src homology 2 domains. Proc Natl Acad Sci USA 90:4251-4255

Zwick E, Wallasch C, Daub H, Ullrich A (1999) Distinct calcium-dependent pathways of epidermal growth factor receptor transactivation and PYK2 tyrosine phosphorylation in PC12 cells. J Biol Chem 274:20989-20996

Endotoxin Tolerance – Mechanisms and Beneficial Effects in Bacterial Infection

M. D. Lehner and T. Hartung[1]

[1]Universität Konstanz, Biochemische Pharmakologie, Fach M655,
78457 Konstanz, Germany. Phone: +49-7531-884116; FAX: +49-7531-884117;
e-mail: Thomas.Hartung@uni-konstanz.de

Contents

Reviews of Physiology, Biochemistry,
and Pharmacology, Vol. 144
© Springer-Verlag Berlin Heidelberg 2002

Endotoxin Tolerance

Endotoxin or lipopolysaccharide (LPS), a glycolipid of the cell membrane of Gram-negative bacteria, is one of the most potent known stimulators of immune responses. The immune system responds to LPS with a systemic production of proinflammatory cytokines which recruit and activate immune cells to eliminate invading pathogens (Mastroeni et al. 1999; Nakano et al. 1992). Although these cytokines are indispensible for the efficient control of growth and dissemination of the pathogen (Langermans and van Furth 1994; Nakano et al. 1992; van Furth et al. 1994), an overshooting inflammatory response is potentially autodestructive and may lead to microcirculatory dysfunction causing tissue damage, shock and eventually death (Beutler et al. 1985; Galanos and Freudenberg 1993). Injection of high dose LPS induces pathological symptoms resembling those of the septic patient (Burrell 1994).

The term "endotoxin tolerance" describes the phenomenon that immune responses and metabolic changes such as fever, inflammation or weight loss as well as lethality in response to LPS challenge are mitigated after repeated LPS administration. Prophylactic subtoxic LPS administration confers protection against inflammatory damage in a number of animal models. Intensive studies attempting to unravel the underlying mechanisms have been conducted over several decades to find a more effective prophylaxis and therapy of Gram-negative infection. In this review data are summarized on two different and apparently contrasting aspects of endotoxin tolerance, i.e. attenuation of inflammatory damage on the one hand and the concomitant modulation of anti-microbial host defense on the other hand.

In vivo Studies

The first reports on acquired resistance to endotoxin derive from physicians, who used vaccines containing whole bacteria to induce fever as a therapeutic measure. In that setting, the development of tolerance to the pyrogenicity of the vaccines was an annoying problem, as it required the infusion of steadily increasing doses to maintain elevated temperatures. Experimentally, Centanni was the first to demonstrate acquired resistance to a purified pyrogenic preparation from bacterial culture filtrate. Repeated injections of rabbits with this heat-stable, non-protein "pyrotoxina bacterica" resulted in a progressive reduction of its fever-inducing activity (Centanni 1894). In 1942, Centanni postulated that the phenomenon was due to a cellular mechanism and not based on a serological immune response (Centanni 1942). Similar results of antibody-independent desensitization to fever induction by repeated administration of fractions from *Salmonella typhosa* to humans were suggested by Favorite and Morgan (Favorite and Morgan 1942). In a set of experiments on pyrogenic tolerance to daily endotoxin infusions in rabbits, Beeson provided further evidence for non-immunologic mechanisms of tolerance. This conclusion was based on the findings that pyrogenic tolerance was not specific for the polysaccharide side chain and could not be transferred to naive animals. Furthermore, tolerance rapidly waned after discontinuation of the daily endotoxin infusions (Beeson 1946; Beeson 1947). Intensive studies performed by Greisman et al. led to the distinction of two phases of endotoxin tolerance (Greisman et al. 1969; Greisman et al. 1966). As reviewed in detail by Johnston et al. (Johnston and Greisman 1985), these are a nonspecific early phase, which becomes evident hours or days after endotoxin treatment and an antibody-dependent late phase tolerance induced by repeated injections of endotoxin. The early phase which lasts for about 48 h until several days is associated with hyporesponsiveness to endotoxins as a class, i.e. tolerance extends to endotoxins unrelated to the one used for desensitization. It is independent of antibody formation, as early tolerance develops equally in athymic (nude) mice, B-cell deficient (xid) mice, and splenectomized mice (Madonna and Vogel 1986). In contrast, several days after endotoxin injection, nonspecific tolerance wanes and hyporesponsiveness is restricted to the endotoxin serotype employed during the pretreatment phase. This late phase tolerance was shown to depend on the formation of LPS-specific antibodies and thus can be passively transferred with serum to naive animals (reviewed by Greisman 1985).

Early reports ascribed the diminished LPS responsiveness after endotoxin pretreatment to increased LPS clearance and degradation, e.g. by stimulation of LPS uptake by the reticuloendothelial system (RES) (Beeson 1946; Beeson 1947). This view was extended by Freedman, who demonstrated that serum transfer of tolerance to the pyrogenic and lethal activities of endotoxin was related to enhanced RES phagocytic activity of recipient rabbits, as assessed by clearance of colloidal carbon (Freedman 1960a; Freedman 1960b). Further studies in contrast demonstrated the development of pyrogenic tolerance in the absence of enhanced phagocytic activity of the RES (Greisman et al. 1964). Moreover, it was shown later on that the administration of thorothrast, used in the early experiments to demonstrate a critical involvement of the RES in mediating LPS tolerance, equally enhanced the fever response in tolerant and naive animals. Hence, a major contribution of enhanced LPS uptake by the RES is considered unlikely. An alternative explanation was suggested by Moreau et al. who demonstrated

Table 1. Endotoxin tolerance *in vivo*

Effect			References
Survival		↑	mouse (Beeson 1946; Freudenberg and Galanos 1988; Hill et al. 1974; Lehner et al. 2001a; Leon et al. 1992;Bundschuh et al. 1997a); rat (Sanchez-Cantu et al. 1989; Wise et al. 1983; Zingarelli et al. 1995b)
fever		↓	human (Astiz et al. 1995; Centanni 1894; Centanni 1942; Favorite and Morgan 1942); guinea pig (Roth et al. 1997; Roth and Zeisberger 1995); rabbit (Dinarello et al. 1968)
TNF	protein	↓	human (Astiz et al. 1995; Kiani et al. 1997; Mackensen et al. 1992); mouse (Balkhy and Heinzel 1999; Baykal et al. 1999; Erroi et al. 1993; Evans and Zuckerman 1991; Faggioni et al. 1995; Gustafson et al. 1995; Lehner et al. 2001a; Leon et al. 1992; Matsuura et al. 1994b; Mengozzi et al. 1991; Schade et al. 1996; Zuckerman and Evans 1992; Zuckerman et al. 1991); guinea pig (Roth et al. 1997); rat (Sanchez-Cantu et al. 1989); rabbit (Wakabayashi et al. 1994); pig (Klosterhalfen et al. 1992)
	mRNA	↓	mouse (Evans and Zuckerman 1991; Zuckerman and Evans 1992); rat (Flohe et al. 1999)
IL-1β	protein	↓	human (Kiani et al. 1997); mouse (Erroi et al. 1993); rabbit (Wakabayashi et al. 1994)
		↑	mouse (Zuckerman et al. 1991)
	mRNA	↓	mouse (Zuckerman and Evans 1992)
IL-6	protein	↓	human (Astiz et al. 1995; Kiani et al. 1997), mouse (Baykal et al. 1999; Erroi et al. 1993; Lehner et al. 2001a; Leon et al. 1992; Mengozzi et al. 1991); guinea pig (Roth et al. 1997)
		↑	human (Mackensen et al. 1992)
IL-8	protein	↓	human (Astiz et al. 1995; Kiani et al. 1997; Mackensen et al. 1992)
IL-10	protein	↓	mouse (Baykal et al. 1999); rat (Flohe et al. 1999)
IL-12	protein	↓	mouse (Balkhy and Heinzel 1999)

Table 1 (continued)

Effect			References
CSF	protein	↓	human (Kiani et al. 1997; Mackensen et al. 1992); mouse (Erroi et al. 1993; Henricson et al. 1990; Henricson et al. 1991; Madonna et al. 1986; Madonna and Vogel 1985; Madonna and Vogel 1986; Quesenberry et al. 1975; Williams et al. 1983)
IFNγ	protein	↓	mouse (Balkhy and Heinzel 1999; Erroi et al. 1993; Gustafson et al. 1995; Lehner et al. 2001a)
chemokines	protein	↓	rat (Blackwell et al. 1997)
arachidonic acid metabolites		↓	rat (Wise et al. 1983); pig (Klosterhalfen et al. 1992)
nitric oxide derivatives		↓	rat (Chamulitrat et al. 1995; Szabo et al. 1994); chicken (Chang et al. 1996)
		=	mouse (Gustafson et al. 1995)
angiotensinogen		↓	rat (Takano et al. 1993)

enhanced activity of serum esterases resulting in increased intravascular inactivation of endotoxin in LPS-pretreated animals (Moreau and Skarnes 1973). Nevertheless, macrophages play a cardinal role in early endotoxin tolerance as demonstrated by Freudenberg et al. in a set of adoptive transfer experiments: in the model of LPS-induced liver injury in galactosamine-sensitized mice Freudenberg et al. revealed that not only LPS toxicity (Freudenberg et al. 1986), but also induction of tolerance required the presence of functional, LPS-sensitive macrophages (Freudenberg and Galanos 1988). Concomitant with the finding that most of the effects of LPS were transmitted by cytokines, several groups reported decreased levels of macrophage derived mediators in endotoxin-tolerant animals (Erroi et al. 1993; Madonna et al. 1986; Mengozzi et al. 1991) and humans (Astiz et al. 1995; Kiani et al. 1997; Mackensen et al. 1992) in response to a second LPS challenge (Table 1). Most studies focused on the production of tumor necrosis factor (TNF), which is almost completely downregulated during LPS tolerance, but other cytokines are also affected by endotoxin pretreatment. Erroi et al. established an order of cytokine inhibition *in vivo* within the same model of LPS tolerance in mice: TNF, interleukin-6 (IL-6) >> colony stimulating factor (CSF) > interferon gamma (IFNγ) > IL-1α and β (Erroi et al. 1993). Downregulation of TNF in spleens and peritoneal macrophages of LPS-tolerant mice appeared already at the level of mRNA production, suggesting a suppression of signaling cascades prior to transcription (Zuckerman and Evans 1992). Whereas downregulation of CSF, IFNγ and IL-6 during LPS tolerance is well established, the effect of repeated LPS injections on IL-1 production is controversial. Several studies showed a partial reduction in circulating IL-1 in response to repeated LPS challenge (Zuckerman and Evans 1992; Zuckerman et al. 1991), whereas in one study IL-1 was even increased (Wakabayashi et al. 1994).

LPS tolerance develops rapidly within several hours, depending on the model. Thus, protection against liver damage of galactosamine-sensitized mice could be induced by LPS injection one hour prior to GalN/LPS challenge (Freudenberg and Galanos 1988). In contrast, suppression of cytokine production took at least five hours after a single dose of LPS (Klosterhalfen et al. 1992). Tolerance to the fever inducing activity of endotoxin even required at least 3 daily injections of endotoxin (Beeson 1947). These kinetic differences suggest distinct mechanisms of LPS-induced protection in the different models, which will be discussed later.

Ex vivo Studies

Further evidence for a contribution of macrophages to LPS tolerance stemmed from *ex vivo* studies showing impaired cytokine production by macrophages isolated from LPS-tolerant animals restimulated *in vitro* (Table 2). In 1968, Dinarello et al. already demonstrated that Kupffer cells isolated from LPS-tolerant rabbits were unable to produce endogenous pyrogen *in vitro* (Dinarello et al. 1968). Peritoneal murine and rat macrophages (resident or thioglycolate-elicited) isolated after *in vivo* administration of LPS displayed a decreased production of TNF (Gahring and Daynes 1986; Haslberger et al. 1988; Moore et al. 1990; Zuckerman et al. 1989) or IL-1 (Gahring and Daynes 1986) upon LPS restimulation *in vitro*. Similarly, impaired production of IL-12 and consequently of IFNγ by spleen cells from endotoxin-tolerant mice was reported (Balkhy and Heinzel 1999). Additionally, these cells displayed decreased responsiveness to substitution with exogenous IL-12, arguing for a suppression of IFNγ production via two distinct mechanisms (Balkhy and Heinzel 1999). Bundschuh et al. demonstrated that suppression of TNF production upon *in vitro* restimulation was a common feature of various macrophage populations (bone marrow cells, peritoneal cells, blood monocytes, alveolar cells and spleen cells) isolated from endotoxin-tolerant mice (Bundschuh et al. 1997b). Monocyte hyporesponsiveness was also reported after administration of endotoxin to humans (Granowitz et al. 1993; Rodrick et al. 1992). However, Mackensen et al. reported an increased capacity to release cytokines upon restimulation *in vitro*

Table 2. Mediator dysregulation in endotoxin tolerance *ex vivo*

Effect			References
TNF	protein	→	human blood (Kimmings et al. 1996), human PBMC (Granowitz et al. 1993); mouse PMΦ (Flach, 1997 #390; Bundschuh, 1997 #288), mouse blood (Schade et al. 1996); other mouse macrophages (Bundschuh et al. 1997a); rat PMΦ (Moore et al. 1990); rat Kupffer cells (Hafenrichter et al. 1994; Hartung and Wendel 1992); rabbit PBMC (Wakabayashi et al. 1994)
		←	human PBMC (Mackensen et al. 1992)
IL-1β	protein	→	human PBMC (Granowitz et al. 1993); mouse PMΦ (Gahring and Daynes 1986); rabbit PBMC (Wakabayashi et al. 1994)
		←	human PBMC (Mackensen et al. 1992)
IL-6	protein	→	human PBMC (Granowitz et al. 1993); rat Kupffer cells (Hafenrichter et al. 1994), rat PMΦ (Zingarelli et al. 1995a)
		←	human PBMC (Mackensen et al. 1992)
IL-10, TGFβ	protein	→	mouse PMΦ (Flach and Schade 1997)
IL-12, IFNγ	protein	→	mouse spleen cells (Balkhy and Heinzel 1999)
arachidonic acid metabolites		→	mouse PMΦ (Haslberger et al. 1988); rat PMΦ (Coffee et al. 1992; Hafenrichter et al. 1994; Moore et al. 1990; Rogers et al. 1986; Rogers et al. 1989; Zingarelli et al. 1995b)
nitric oxide derivatives		←	rat PMΦ (Zingarelli et al. 1995a; Zingarelli et al. 1995b)
superoxide		→	rat non parenchymal cells, perfused liver (Bautista and Spitzer 1995)

of PBMC from endotoxin pretreated cancer patients, although serum cytokine levels were significantly reduced after repeated LPS injection. In contrast to the other two studies with human volunteers, Mackensen et al. isolated PBMC from cancer patients 24 hours after the last LPS injection, whereas in the other studies blood was withdrawn one hour or 6 hours, respectively, after LPS injection (Mackensen et al. 1992).

In vitro Studies

Most studies on the mechanism of macrophage desensitisation derive from experiments using primary cells or immortalized cell lines exposed to repeated LPS stimuli *in vitro* (Cavaillon et al. 1994; Matic and Simon 1991). As shown for macrophages isolated from endotoxin-tolerant hosts, release of various macrophage mediators in response to LPS stimulation is mitigated after repeated exposure to endotoxin *in vitro*. In this review we will refer to this status of macrophage hyporesponsiveness induced by repeated LPS stimulation *in vitro* as macrophage desensitization or refractoriness, to differentiate it from *in vivo* LPS tolerance, which might involve other mechanisms additional to downregulation of cytokine production. Suppression of cytokine release after LPS exposure was demonstrated for primary cells, such as peritoneal macrophages from mouse or rabbit and human monocytes as well as a variety of murine and human cell lines (Table 3). The spectrum of cytokines downregulated in desensitized macrophages *in vitro* involves the same mediators shown to be suppressed *in vivo*, although controversial data were provided for most cytokines except TNF. Thus, depending on the experimental setting, downregulation of TNF after exposure to endotoxin was associated with unchanged production as well as suppression or increase of IL-1, IL-6, IL-8, IL-10, and PGE_2 release in response to a subsequent LPS stimulus (Frankenberger et al. 1995; Lepe-Zuniga and Klostergaard 1990; Mengozzi et al. 1993; Randow et al. 1995; Seatter et al. 1995; Takasuka et al. 1991). Most controversial data were obtained on the regulation of IL-1. Whereas studies performed with the human cell line THP-1 revealed a downregulation of IL-1 mRNA and protein after repeated LPS stimulation (LaRue and McCall 1994; Yoza et al. 1998; Yoza et al. 2000), data derived from experiments using human or mouse primary cells demonstrated unchanged or even increased IL-1 production in response to a second LPS stimulus (Heagy et al. 2000; Knopf et al. 1994; Kraatz et al. 1998; Li et al. 1994; Mengozzi et al. 1993; Takasuka et al. 1995; Takasuka et al. 1991). However, it is possible that despite normal or increased IL-1 levels in desensitized macrophages, the biological activity of IL-1 in the supernatant is

Table 3. Mediator dysregulation in endotoxin desensitization in vitro

Effect			References
TNF	protein	→	human PMΦ (Knopf et al. 1994) human monocytes (Annenkov and Baranova 1991; Cavaillon et al. 1994; Karp et al. 1998; Lepe-Zuniga and Klostergaard 1990; Matic and Simon 1991; Mengozzi et al. 1993; human PBMC (Heagy et al. 2000; Randow et al. 1997; Randow et al. 1995; Riedel and Kaufmann 2000); human dendritic cells (Karp et al. 1998); mouse RAW 264.7 cell line (Virca et al. 1989); mouse PMΦ (Ancuta et al. 1997; Kraatz et al. 1998; Kraatz et al. 1999a; Kraatz et al. 1999b; Li et al. 1994; Matsuura et al. 1994b; Sato et al. 2000; Takasuka et al. 1991; West et al. 1997; rabbit PMΦ (Mathison et al. 1990)
	mRNA	→	human MonoMac 6 cell line (Frankenberger et al. 1995; Kastenbauer and Ziegler-Heitbrock 1999; Labeta et al. 1993; Ziegler-Heitbrock et al. 1995); human PMΦ (Knopf et al. 1994); mouse RAW 264.7 cell line (Virca et al. 1989); mouse PMΦ (Bohuslav et al. 1998; Matsuura et al. 1994b; Takasuka et al. 1995; Takasuka et al. 1991; Tominaga et al. 1999)
IL-1β	protein	→	human THP-1 cell line (LaRue and McCall 1994; Yoza et al. 1998)
		=	human monocytes (Mengozzi et al. 1993); mouse PMΦ (Takasuka et al. 1991)
		←	human PBMC (Heagy et al. 2000); human PMΦ (Knopf et al. 1994); mouse PMΦ (Kraatz et al. 1998; Li et al. 1994)
	mRNA	→	human THP-1 cell line (LaRue and McCall 1994; Yoza et al. 1998; Yoza et al. 2000)
		=	mouse PMΦ (Takasuka et al. 1995; Takasuka et al. 1991)
IL-6	protein	→	human PMΦ (Knopf et al. 1994); human PBMC (Riedel and Kaufmann 2000); human monocytes (Shimauchi et al. 1999); mouse PMΦ [Tominaga, 1999 #888; Nomura, 2000 #863]
		=	human monocytes (Mengozzi et al. 1993) (Wittmann et al. 1999); mouse PMΦ [Li, 1994 #75]

Table 3 (continued)

Effect			References
IL-10	protein	→	human PBMC (Riedel and Kaufmann 2000; Randow et al. 1997; Randow et al. 1995)
		←	human MonoMac 6 cell line (Frankenberger et al. 1995); human monocytes (Shimauchi et al. 1999)
IL-12	protein	→	human monocytes (Karp et al. 1998), human dendritic cells (Karp et al. 1998)
IL-1ra	protein	←	human THP-1 cell line (Learn et al. 2000)
		=	human PBMC (Randow et al. 1995)
CSF	protein	→	human PBMC (Riedel and Kaufmann 2000); mouse PMΦ (Medvedev, 2000 #855)
		←	human PMΦ (Knopf et al. 1994)
arachidonic acid metabolites		→	mouse PMΦ (Li et al. 1994)
nitric oxide derivatives		→	mouse PMΦ (Bogdan et al. 1993; Hirohashi and Morrison 1996; Severn et al. 1993; Zhang and Morrison 1993); chicken macrophages (Chang et al. 1996)
		←	mouse PMΦ (Ancuta et al. 1997; West et al. 1994; Zhang and Morrison 1993)
MHC II expression		→	human PBMC (Wolk et al. 2000)

suppressed due to sustained or even increased expression of the natural antagonist IL-1 receptor antagonist (IL-1ra) (Learn et al. 2000; Randow et al. 1995).

Suppression of TNF release was associated with decreased mRNA levels, suggesting transcriptional control of cytokine production in cell lines (Frankenberger et al. 1995; Kastenbauer and Ziegler-Heitbrock 1999; Labeta et al. 1993; Virca et al. 1989; Ziegler-Heitbrock et al. 1995), human (Knopf et al. 1994), mouse (Bohuslav et al. 1998; Matsuura et al. 1994b; Takasuka et al. 1995; Takasuka et al. 1991; Tominaga et al. 1999), and rabbit primary cells (Mathison et al. 1990). This view was challenged by Zuckerman et al. demonstrating inhibition of TNF release despite increased mRNA levels in LPS-pretreated cells (Zuckerman et al. 1989).

Besides cytokine production, the regulation of nitric oxide (NO) synthesis during LPS tolerance has been studied in detail, but the results are as controversial as for regulation of IL-1. Expression of inducible NO synthase (iNOS) and NO production in response to a second LPS stimulus were suppressed (Bogdan et al. 1993; Chang et al. 1996; Hirohashi and Morrison 1996) or increased (Ancuta et al. 1997; West et al. 1994), depending on the experimental settings. In line with these data, Zhang et al. demonstrated that depending on the concentration of the primary LPS stimulus, either suppression or priming of NO production can be found (Zhang and Morrison 1993).

To sum up, *in vitro* exposure of cells to LPS results in suppression of TNF release and reprogrammed production of various other macrophage mediators in response to subsequent stimulation. Cells desensitized *in vitro* display many features of macrophages isolated from endotoxin-tolerant hosts. Despite the apparent limitations of the *in vitro* setting such as neglection of neuroendocrine regulation, glucocorticoids and the interaction of different cell types *in vivo*, much of our current knowledge concerning the mechanisms of macrophage desensitization is derived from *in vitro* experiments.

Mechanisms of Macrophage Desensitization

In the last years, our understanding of the molecular mechanisms underlying desensitization of macrophages by exposure to LPS has increased considerably. Although Larsen et al. suggested that LPS preexposure decreased the number of LPS binding sites on monocytes (Larsen and Sullivan 1984), the expression of the LPS receptor CD14 is unaffected or even increased following LPS-stimulation (Labeta et al. 1993; Matsuura et al. 1994a; Ziegler-Heitbrock et al. 1994). Thus, it is highly unlikely that tolerance is mediated via expression of this LPS receptor. However, recent results by Nomura demonstrated downregulated surface expression of the LPS signaling receptor toll like receptor 4 (TLR4) on LPS-desensitized macrophages (Nomura et al. 2000). Further downstream, refractoriness in response to LPS preexposure has been shown to be associated with altered G-protein content (Coffee et al. 1992; Makhlouf et al. 1998a), phospholipase D and phosphatidylinositol-3 kinase expression (Bowling et al. 1995). West et al. reported compromised protein kinase C (PKC) activation in LPS desensitized cells (West et al. 1997) and receptor independent stimulation of the PKC by phorbol myristate acetate could overcome the suppression of cytokine production associated with refractoriness. Others described suppressed signal transduction via both the mitogen-activated-protein (MAP) kinase cascade (Kraatz et al. 1999a; Kraatz et al. 1999b; Medvedev et al. 2000; Tominaga et al. 1999) and inhibitor of NF-κB (I-κB) kinases, resulting in impaired transcription of nuclear-factor-kappa B (NF-κB)- and activation protein-1 (Ap-1)-regulated genes (Medvedev et al. 2000; Yoza et al. 1998). An alternative mechanism for suppression of NF-κB-dependent gene expression was suggested by Ziegler-Heitbrock et al.. They used a human monocyte cell line (MonoMac-6) to demonstrate an upregulation of the p50 subunit of NF-κB in LPS refractory cells, leading to a predominance of transactivation-inactive p50/p50 homodimers. These homodimers bind to NF-κB motifs in several promotors and thereby inhibit the transcription of genes such as TNF (Kastenbauer and Ziegler-Heitbrock 1999; Ziegler-Heitbrock et al. 1994). Support for this hypothesis originates from experiments with p50 deficient mice that are resistant to tolerance induction by LPS (Bohuslav et al. 1998). Inhibition of gene transcription in response to a second LPS stimulus via the formation of a specific nuclear suppressor of LPS-induced gene transcription was also suggested by others (LaRue and McCall 1994; Yoza et al. 2000). LaRue et al. provided evidence that decreased LPS-induced transcription of IL-1β in LPS-desensitized THP-1 cells was regulated by a labile repressor which required constant protein synthesis and suggested IκB-α as a potential candidate, although then a contribution of p50 had not yet been studied (LaRue and McCall 1994).

In contrast, recent data showing decreased surface expression of TLR4 on LPS-tolerized cells (Nomura et al. 2000) and suppression of IL-1 receptor-associated kinase (IRAK) activation and association with myeloid differentiation protein (MyD88) (Li et al. 2000), support the notion that already very early steps in LPS-signaling upstream of NF-κB are altered after LPS exposure. Further evidence for this was provided by Medvedev et al. (Medvedev et al. 2000) who re-evaluated *in vitro* desensitization by IL-1 and TNF, showing induction of cross-tolerance to LPS via the IL-1 receptor but not the TNF-receptor. Intriguingly, signal transduction of the IL-1R, the LPS-receptor TLR4, and TLR2 employ similar signaling molecules (Medzhitov et

al. 1998; Yang et al. 1998; Zhang et al. 1999). Recent studies from our laboratory demonstrated that preexposure to lipoteichoic acid that induced signaling via TLR2 resulted in hyporesponsiveness to TLR4-mediated LPS signaling and vice versa (Lehner et al. 2001b). This finding adds further indirect
evidence for a suppression of common signaling molecules shared by
TLR2/4 and IL-1R, i.e. MyD88, IRAK, TNF receptor-activated factor 6
(TRAF6) or NF-κB-inducing kinase (NIK) in desensitized macrophages. The
view that inhibition of common signaling pathways of the IL-1R/TLR family
and not diminished TLR4 surface expression is mainly responsible for
macrophage hyporesponsiveness is corroborated by the finding that preexposure of macrophages to the TLR2-dependent stimulus mycoplasmal
lipopeptide (MALP-2) suppressed LPS-induced TNF release without reducing the surface expression of TLR4 (Sato et al. 2000).

Despite the large number of studies dealing with macrophage hyporesponsiveness in response to LPS pretreatment, the exact mechanism of suppression of cytokine production has not been identified yet. Since there is
sound evidence for a contribution of various of the aforementioned factors,
it is feasible that i) macrophage desensitization is the result of the orches-
trated action of multiple factors activated by the primary LPS stimulus or ii)
depending on the model employed to study tolerance (species, cell type,
experimental settings) varying distinct mechanisms account for refractoriness in response to inflammatory bacterial components.

Mediators of Tolerance

LPS exerts most of its effects via the activity of macrophage mediators released in response to LPS stimulation. The inflammatory response is regulated by a complex network of mediators that directly interact with each
other's expression or biological activity. In this context, a number of macrophage mediators such as IL-10, TGFβ or PGE_2 have potent anti-inflammatory activity by suppressing the formation of proinflammatory cytokines
(Berg et al. 1995; de Waal Malefyt et al. 1991; Howard et al. 1993; Renz et al.
1988; Scales et al. 1989; Smith et al. 1996). Thus, it has been presumed that
autocrine mechanisms are also involved in suppression of cytokine production during LPS tolerance.

Mediators of *in vitro* Desensitization

As outlined before, LPS-pretreatment of cultured macrophages results in
hyporesponsiveness to cytokine release in response to a subsequent LPS
stimulus. It has been shown that several cytokines could substitute for LPS

as the desensitizing stimulus. Cavaillon et al. demonstrated that incubation of human PBMC with recombinant cytokines prior to restimulation with LPS partially suppressed production of TNF to a different extent. Whereas preexposure to TGFβ or IL-10 reduced TNF release by nearly 60% as compared to saline pretreated cells, IL-4 and IL-1 were less effective (35% and 30% inhibition, respectively) and no inhibition at all was found after administration of TNF, IL-6, IL-8 or leukaemia inhibitory factor (Cavaillon et al. 1994). The lack of macrophage desensitization by exposure to TNF has been reported also by Li et al. (Li et al. 1994). The differential role of TNF and IL-1β in desensitization of macrophages *in vitro* was confirmed by recent studies from Medvedev et al. who showed that exposure of murine macrophages to LPS or IL-1, but not to TNF, resulted in inhibition of transcription factor activation and suppressed transcription of GM-CSF and several chemokines in response to a second LPS stimulus (Medvedev et al. 2000). Unfortunately, no information on the regulation of TNF mRNA and protein was given in this study. Convincing data on the contribution of soluble mediators in desensitization of macrophages were derived from experiments with human PBMC. Randow et al. demonstrated that a combination of recombinant human IL-10 and TGFβ was as effective as low-dose LPS pretreatment in terms of reduction of TNF release upon subsequent high-dose LPS stimulation, whereas preexposure to either cytokine alone only partially suppressed the release of TNF (Randow et al. 1995). In the same setting, addition of neutralizing antibodies to IL-10 and TGFβ inhibited desensitization in response to the first LPS stimulus, providing direct proof for a contribution of these two anti-inflammatory cytokines in LPS-induced monocyte/macrophage refractoriness *in vitro* (Randow et al. 1995). The critical role of IL-10 and TGFβ in downregulation of TNF production was confirmed by Karp et al., whereas inhibition of IL-12 production in LPS-pretreated human monocytes was independent of these cytokines (Karp et al. 1998). In line, antibodies against IL-4 or IL-10 as well as addition of indomethacin or a iNOS inhibitor did not abrogate suppression of IL-12 p40 mRNA and protein expression in LPS-desensitized macrophages (Wittmann et al. 1999).

The production of a yet unidentified suppressor of TNF formation not identical with IL-1, IL-10 or TGFβ during endotoxin tolerance was reported by Schade et al (Flach and Schade 1997; Schade et al. 1996). They showed that addition of culture supernatants of LPS-stimulated peritoneal murine macrophages from endotoxin pretreated mice suppressed TNF release by naive macrophages. Similar results on a selective inhibitor of TNF, but not of IL-1 or IL-6 synthesis in supernatants of LPS-desensitized macrophages were provided by Fahmi et al. (Fahmi and Chaby 1993). The idea of a negatively acting autocrine mediator in macrophage desensitization was extended by more recent results from Baer et al. who demonstrated the production of a yet unidentified "TNF-inhibiting factor" (TIF) in supernatants of a LPS-stimulated macrophage cell line. Inhibition of TNF expression by macrophage conditioned medium was associated with selective induction of the NF-κB p50 subunit which selectively inhibited a TNF-promoter reporter construct (Baer et al. 1998). Since a contribution of IL-4, IL-10 and TGFβ was excluded, these findings provide evidence for LPS induction of a novel cytokine with selective TNF-inhibitory potential participating in endotoxin desensitization (Baer et al. 1998).

Besides cytokines, arachidonic metabolites were shown to influence the responsiveness of macrophages. It is well established that prostaglandin E_2 (PGE_2) downregulates TNF production by macrophages, probably via the elevation of cAMP (Kunkel et al. 1988; Renz et al. 1988; Scales et al. 1989). Thus it is feasible that PGE_2 produced in response to the primary desensitizing dose of LPS contributes to macrophage hyporesponsiveness. This view was supported by the finding that PGE_2 production was increased in LPS-desensitized macrophages (Li et al. 1994; Makhlouf et al. 1998b; Matic and Simon 1991; Seatter et al. 1995). However, direct addition of PGE_2 during primary culture failed to suppress TNF production upon subsequent LPS stimulation of cultured human monocytes (Matic and Simon 1991). In addition, in three different studies the addition of the cyclooxygenase inhibitor indomethacin neither prevented the development of hyporesponsiveness nor restored TNF production upon LPS restimulation (Mathison et al. 1990; Takasuka et al. 1991; Ziegler-Heitbrock et al. 1992). In contrast, by using higher concentrations of indomethacin (10–100 μM), Haas et al. could inhibit the suppression of TNF production by LPS-pretreatment (Haas et al. 1990). Thus, the contribution of arachidonic acid derivatives in desensitization of macrophages still remains to be clarified. Our recent results derived from co-culture experiments argue against a major role of soluble mediators in acquired hyporesponsiveness. Cross-desensitization induced by pre-exposure to LPS or LTA in wild-type macrophages was not transferred to co-cultured macrophages from mice lacking functional TLR2 or TLR4 as evidenced by sustained TNF release upon re-challenge with the other stimulus (Lehner et al. 2001b). However, as we did not perform any neutralization experiments, we cannot rule out that, besides ligand-TLR interaction additional signals provided by soluble mediators were required for desensitization.

Mediators of *in vivo* Tolerance

As pointed out for macrophage desensitization *in vitro*, the involvement of soluble mediators in establishing LPS tolerance *in vivo* has also been discussed controversially. Attempts to induce tolerance to the pyrogenicity of subsequent endotoxin injection by repeated administration of endogenous pyrogen (EP) were not successful (Atkins 1960). In contrast, pretreatment of rabbits with IL-1 partially abolished hypotension and TNF release in response to subsequent endotoxin challenge (Kaplan et al. 1993). When mice were treated with recombinant TNF or IL-1α, neither cytokine alone was able to mimic LPS induction of tolerance. However, the two cytokines synergized to induce features of early endotoxin tolerance, such as alterations of the monocyte/macrophage bone marrow pool and suppression of CSF release upon subsequent LPS challenge (Vogel et al. 1988). In addition, suppression of CSF release associated with LPS-tolerance was partially reversed by administration of recombinant IL-1 receptor antagonist (IL-1ra) during LPS pretreatment (Henricson et al. 1991). Administration of IL-1α or TNF but not of IL-6 to mice for four days partially inhibited the production of IL-6 and TNF in response to a subsequent LPS challenge, although to a lesser extent than LPS (Erroi et al. 1993). In line with this finding, TNF infusion in rats resulted in a reduced capacity of isolated bone marrow cells to produce TNF, IL-6 or PGE_2 upon LPS stimulation *in vitro* (Ogle et al. 1997). In contrast, Mathison et al. failed to suppress the production of TNF in response to LPS by pretreating rabbits with TNF infusions (Mathison et al. 1988). Pretreatment with IL-1 conferred protection to subsequent high dose LPS challenge (Alexander et al. 1991b; Leon et al. 1992) and sepsis induced by cecal ligation and puncture (CLP) (Alexander et al. 1991a) as well as *E. coli* induced peritonitis (Lange et al. 1992). Similar results were obtained for TNF, which induced tolerance to the lethality of subsequent LPS challenge (Alexander et al. 1991b; Fraker et al. 1988). In the model of inflammatory liver damage in galactosamine (GalN)-sensitized mice pretreatment with TNF or IL-1 was equally protective as LPS in reducing the extent of liver damage and lethality (Bohlinger et al. 1995; Libert et al. 1991; Wallach et al. 1988). Moreover, administration of IL-1, TNF or LPS induced tolerance to the toxicity of TNF injection itself, as shown for the metabolic changes, weight loss, temperature increase and lethality in response to high-dose TNF injection (only TNF pretreatment) (Fraker et al. 1988; Roth and Zeisberger 1995), as well as for low-dose TNF-induced hepatocyte apoptosis in GalN-sensitized mice (TNF or IL-1 pretreatment) (Bohlinger et al. 1995; Libert et al. 1991; Wallach et al. 1988). Since enhanced clearance or neutralization of TNF in LPS- or cytokine-pretreated animals was excluded (Fraker et al. 1988;

Roth and Zeisberger 1995) hyporesponsiveness of target cells to TNF activity itself, e.g. by downregulation of TNF receptors and by the production of acute phase proteins or anti-apoptotic factors was suggested as an additional mechanism contributing to LPS tolerance (Bohlinger et al. 1995; Hartung and Wendel 1992; Libert et al. 1991; Wallach et al. 1988). This view was corroborated by the finding that addition of acute phase proteins attenuated the GalN/TNF induced liver damage (Libert et al. 1996; Van Molle et al. 1999; Van Molle et al. 1997). Thus, the protection afforded by LPS pretreatment in the GalN/LPS model is likely to be mediated by two independent mechanisms differing in their requirement of endogenously produced cytokines. On the one hand, the reduction of TNF levels in mice pretreated with LPS suggests macrophage hyporesponsiveness similar to *in vitro* desensitization. As discussed before, although a role of soluble mediators in macrophage desensitization *in vitro* has not been fully identified, yet, evidence has been provided that soluble mediators do not suffice for downregulation of macrophage responsiveness. This view is substantiated by our unpublished results showing suppression of TNF release in TNFR1 deficient mice in response to repeated LPS injections. On the other hand, it is likely that TNF and IL-1 produced upon LPS pretreatment induce hyporesponsiveness of hepatocytes to TNF activity itself as an additional mechanism of protection.

As outlined before, several *in vitro* studies suggested that LPS-induced desensitization of macrophages was mediated via formation of IL-10. In line with this, administration of IL-10 protected mice against a lethal endotoxin challenge (Howard et al. 1993). However, a major role of the antiinflammatory cytokine IL-10 in mediating LPS tolerance *in vivo* was excluded by Berg et al. using IL-10 deficient mice. Although these mice were LPS-hyperresponsive in terms of TNF production and lethality, tolerance after an initial sublethal LPS dose developed normally as determined by decreased lethality and diminished levels of TNF and IL-6 after subsequent high dose LPS challenge. In addition, infusion of recombinant IL-10 could not substitute for the initial desensitizing dose of LPS (Berg et al. 1995). In conclusion, although evidence has been provided that cytokines such as TNF or IL-1 have the potential to mimic some of the beneficial effects of LPS pretreatment *in vivo* the actual role of these cytokines in LPS-induced macrophage desensitization still has to be characterized. One important point is that most investigators used recombinant cytokines produced in *E. coli*. Since a possible endotoxin contamination of these recombinant cytokines had not always been excluded, it is difficult to ascribe the observed effects of recombinant proteins to cytokine activity.

Besides cytokines, glucocorticoids possess a strong anti-inflammatory potential. Administration of cortisone prevented lethality after high dose

LPS challenge (Geller et al. 1954) and suppressed the release of TNF, IL-1 and IL-6 (Beutler et al. 1986; Coelho et al. 1995; Goujon et al. 1996; Morrow et al. 1993; Steer et al. 2000; Waage and Bakke 1988). In line, adrenalectomy sensitized mice to the toxicity of subsequent LPS injection (Evans and Zuckerman 1991; Goujon et al. 1996; Parant and Chedid 1971). Moreover, since glucocorticoids are released in response to LPS injection, it was feasible to ascribe endotoxin tolerance to the anti-inflammatory activity of endogenous glucocorticoids (Evans and Zuckerman 1991). Studies by Evans demonstrated that LPS tolerance could not be induced in adrenalectomized mice (Evans and Zuckerman 1991). However, this view was challenged by the finding that endotoxin tolerance in terms of suppressed TNF release developed normally in adrenalectomized rats (Chautard et al. 1999). This finding confirmed earlier results from Chedid's group. In their experiments, endotoxin tolerance developed equally in the absence of glucocorticoids, as shown by adrenalectomy prior to or directly after the initial desensitizing injection of LPS, albeit on the background of overall increased susceptibility (Chedid et al. 1964). A similar status of LPS-hyperresponsiveness can be induced by repeated injections of cortisone. Also under this condition of decreased glucocorticoid responsiveness, mice were rendered endotoxin-tolerant by a single LPS injection (Chedid et al. 1964). Studies in the GalN/TNF model demonstrated that addition of dexamethasone did not prevent liver injury (Libert et al. 1991), indicating that at least one aspect of LPS tolerance, i.e. diminished sensitivity of hepatocytes to cytokine activity was not mediated by glucocorticoids.

These results, together with the finding that suppression of cytokine release can also be induced *in vitro* (i.e. in the absence of glucocorticoids), argue against a critical involvement of glucocorticoids in endotoxin tolerance.

Specificity of Tolerance

The question, whether early phase nonspecific tolerance is restricted to endotoxins as a class or whether it reflects a general state of altered macrophage activity resulting in diminished cytokine expression in response to non-endotoxin inflammatory stimuli as well, has not been settled. The view that tolerance is restricted to endotoxins as a class originates from experiments performed by Greisman et al. who demonstrated that rabbits rendered LPS-tolerant by infusion of endotoxin for several hours displayed a normal fever reaction in response to pyrogenic non-endotoxin challenges such as influenza virus, old tuberculin and staphylococcal enterotoxin (Greisman et al. 1966). Similarly, Roth et al. showed a lack of cross-reactivity between LPS and muramyl-dipeptides in terms of fever induction and production of TNF and IL-6 in guinea pigs (Roth et al. 1997). However, the experimental setting used consisting of repeated injections of endotoxin over a period of 15 days with administration of muramyl-dipeptide 3 days after the last LPS injection may have been unsuitable to study the specificity of the early phase tolerance which is most prominent within the first 48 h and then starts to wane. Lack of cross-tolerance was reported also by Mathison et al. who failed to suppress TNF-release in response to *Staphylococcus aureus* by preexposure of rabbit macrophages to LPS (Mathison et al. 1990). Similarly, LPS-tolerant Kupffer cells still produced TNF upon viral infection (Busam et al. 1991). However, differential suppression of TNF and IL-1 was reported by Wakabayashi et al. who showed that PBMC isolated from LPS-tolerant rabbits still produced TNF, but no IL-1, in response to *Staphylococcus epidermidis* (Wakabayashi et al. 1994), proposing differential regulation of these cytokines during hyporesponsiveness.

Further evidence that downregulation of monocyte/macrophage function after LPS-pretreatment is not restricted to restimulation with endotoxins was provided by Granowitz et al. (Granowitz et al. 1993). They demonstrated a reduction of cytokine release by human PBMC derived from endotoxin pretreated volunteers restimulated *ex vivo* with LPS, IL-1 or TSST-1. Cavaillon et al. reported suppression of TNF-release in response to zymosan, staphylococci and streptococci after exposure of human monocytes to LPS *in vitro* (Cavaillon et al. 1993; Cavaillon et al. 1994). Similar results were obtained more recently by Karp et al. for downregulation of IL-12 production (Karp et al. 1998). Further support for a general macrophage desensitization not restricted to LPS stemmed from Kreutz et al., who reported TNF suppression upon repeated exposure to whole *S. aureus* or synthetic lipopeptides (Kreutz et al. 1997). Recently, we could demonstrate macrophage cross-desensitization in terms of TNF production by LPS and lipoteichoic acids (LTA) from *S. aureus* via different TLR (Lehner et al. 2001b). The same held true for *in vivo* tolerance to liver damage by administration of galactosamine plus LPS or LTA. This extends recent findings by Sato et al., who reported cross-desensitization of macrophages by mycoplasmal lipopeptides and LPS via TLR2 and TLR4 (Sato et al. 2000). These results suggest that tolerance and macrophage desensitization could represent a general antiinflammatory mechanism induced by selected bacterial stimuli to prevent potentially harmful overshooting inflammation during sustained infection.

Restoration of Cytokine Response

To study the mechanism of cytokine suppression in LPS-desensitized macrophages, a variety of substances was tested for their ability to overcome suppression of cytokine release in response to a second LPS challenge. Several reports indicated that direct stimulation of protein kinase C by addition of PMA to desensitized macrophages had some potential to restore normal immune functions: In human monocytes pretreated with LPS, TNF release in response to PMA was even increased compared to cells preexposed only to medium (Matic and Simon 1991). Others demonstrated reversal of TNF suppression in desensitized murine macrophages by addition of PMA one hour prior to second LPS activation (West et al. 1997). The restoration of LPS responsiveness by preincubation with PMA was associated with reversed inhibition of MAPK and p38 kinase activation (Kraatz et al. 1999b). In endotoxin-tolerant mice, injection of PMA 10 min before secondary LPS challenge counteracted suppression of IL-6 and partially of CSF production, but had no effect on TNF release while IL-1β production was even down-regulated (Erroi et al. 1993; Mengozzi et al. 1991). More physiological tools to restore cytokine release include the proinflammatory cytokines interferon gamma (IFNγ), granulocyte-macrophage colony stimulating factor (GM-CSF) and interleukin 12 (IL-12): It is well established that IFNγ, produced by T lymphocytes or NK cells upon various inflammatory stimuli, is a potent activator of macrophage functions. These include an upregulation of MHC II expression, enhanced microbicidal activity against intracellular pathogens and release of proinflammatory cytokines (Kagaya et al. 1989; Ozmen 1994), reviewed in (Young and Hardy 1995). IFNγ receptor-deficient mice display decreased sensitivity to LPS toxicity, associated with depressed TNF synthesis, diminished expression of CD14, and low plasma LPS-binding capacity (Car et al. 1994). This suggests that IFNγ is an important co-stimulus for macrophage gene expression that might overcome hyporesponsiveness. Indeed, addition of IFNγ to LPS-desensitized macrophages prior to or concomitant with LPS restimulation partially reversed suppression of TNF production. This was demonstrated for the MonoMac-6 cell line (Haas et al. 1990), human monocyte-derived macrophages (Matic and Simon 1992), human PBMC (Randow et al. 1997) and mouse macrophages isolated after induction of endotoxin tolerance *in vivo* (Bundschuh et al. 1997b). Also several features of *in vivo* endotoxin tolerance, such as suppression of TNF (Bundschuh et al. 1997b; Mengozzi et al. 1991) or IL-6 release (Mengozzi et al. 1991) and increased resistance to endotoxic shock (Bundschuh et al. 1997b), could be partially abolished when IFNγ was injected additionally to LPS re-challenge. Administration of IFNγ also reversed the suppression of

TNF, IL-6 and G-CSF release in LPS-tolerant cancer patients (Mackensen et al. 1991).

Like IFNγ, GM-CSF is involved in regulation of LPS-induced cytokine production and lethality (Basu et al. 1997; Tiegs et al. 1994). Further, addition of GM-CSF partially counteracted macrophage desensitization (Bundschuh et al. 1997a; Randow et al. 1997), but the priming efficacy differed compared to IFNγ, depending on the cell type used. Thus, GM-CSF was more effective than IFNγ in restoring TNF production by murine monocytes and bone-marrow cells, but less effective when more differentiated macrophages such as peritoneal cells or alveolar macrophages were used (Bundschuh 1997). In contrast, suppression of TNF production in LPS-pretreated human PBMC and bone marrow cells was counteracted more efficiently by IFNγ (Bundschuh 1997; Randow et al. 1997). Reversal of TNF suppression in LPS-desensitized human PBMC by addition of IL-12 was also demonstrated recently. This cytokine is normally produced by monocytes/macrophages upon inflammatory stimuli and induces IFNγ release by T lymphocytes and NK cells. Since IL-12 release was shown to be downregulated in desensitized macrophages, substitution with exogenous IL-12 should restore TNF release via production of IFNγ. Direct proof for this hypothesis was provided by Randow et al. who showed that the effect of IL-12 was dependent on both the presence of nonmonocytic cells and production of IFNγ (Randow et al. 1997). However, suppression of IFNγ production by spleen cells from LPS-tolerant mice could only be partially reversed by addition of IL-12, suggesting diminished responsiveness to IL-12 as an additional mechanism of tolerance (Balkhy and Heinzel 1999).

The mechanism underlying restoration of TNF release in LPS-desensitized cells by pretreatment or coadministration of these proinflammatory cytokines is not fully clear yet. Enhancement of TNF production in response to IFNγ, GM-CSF or IL-12 is not restricted to LPS-desensitized cells, but is also found in naive monocytes/macrophages (Bundschuh et al. 1997a; Randow et al. 1997). Thus it is feasible that instead of specifically restoring signaling pathways suppressed during hyporesponsiveness, these cytokines rather act by amplifying the minimal responses that still occur in desensitized cells, using the same pathways involved in the enhancement of primary LPS responses.

Endotoxin Tolerance and Infection

Introduction

Dysbalanced production of leukocyte-derived inflammatory mediators such as cytokines, arachidonic acid metabolites, lysosomal enzymes, reactive oxygen or nitrogen intermediates is considered a major mechanism responsible for pathophysiological alterations of the microcirculation, leading to shock, multiple organ failure and eventually death in response to systemic infection or endotoxaemia (Bone 1996; Bone et al. 1997; Dinarello 1997; Dinarello 2000; Marik and Varon 1998). Experimentally induced endotoxin tolerance provides protection against lethality and morbidity in animal models of endotoxic shock and fulminant infection used to simulate the systemic inflammatory response syndrome of the septic patient. As pointed out before, LPS tolerance is associated with suppression of several cytokines, attenuation of leukocyte infiltration and consequently a reduction of organ damage. These findings suggested induction of LPS tolerance to be an interesting tool in sepsis prophylaxis (Cavaillon 1995; Gustafson et al. 1995; Salkowski et al. 1998). However, concern was raised whether suppression of inflammatory responses during LPS tolerance would interfere with normal host defense and thus predispose patients to nosocomial infection (Cavaillon 1995). Indeed, host defense against infection with small numbers of replicating pathogens requires an intact cytokine response to halt proliferation and dissemination of the pathogen (Mastroeni et al. 1999; Nakano et al. 1992). In contrast to models of acute hyperinflammation such as endotoxic shock, neutralization of proinflammatory cytokines worsens the outcome of infection with low numbers of virulent bacteria (Dai et al. 1997; Tite et al. 1991) and many cytokine-deficient mice that are resistant to inflammatory damage rapidly succumb to otherwise sublethal infections (Dai et al. 1997; Irikura et al. 1999; Mastroeni et al. 1999; O'Brien et al. 1999; Pfeffer et al. 1993; Rothe et al. 1993). Also, mice inherently hyporesponsive to LPS because of a nonfunctional mutation in the tlr4 gene (Hoshino et al. 1999; Poltorak et al. 1998; Qureshi et al. 1999; Vogel et al. 1999) display increased susceptibility to Gram-negative pathogens (Cross et al. 1995; Hagberg et al. 1984). Furthermore, experimentally induced endotoxin tolerance displays features of immunoparalysis observed frequently in post-septic or post-traumatic patients several days or weeks after systemic inflammation: Monocytes from immunoparalysed patients were impaired in their ability to produce TNF upon restimulation with LPS *in vitro* (Docke et al. 1997; Ertel et al. 1995; Kox et al. 1997; McCall et al. 1993; Munoz et al. 1991; Volk et al. 1996; Volk 1991; Wilson et al. 1997) and displayed diminished surface expression of MHC II (Docke et al. 1997; Kox et al. 1997; Volk et al. 1996; Volk 1991). These cellular defects were associated with an increased incidence of infectious complications and lethal outcome of disease (Hershman et al. 1990; Volk 1991). Since similar alterations of monocyte/macrophage activity were found during experimentally induced LPS tolerance, it was feasible that induction of LPS tolerance equally interfered with host defense.

In contrast, it has been known for a long time that endotoxin is a potent activator of host defense and LPS treatment is associated with protection against the lethality of irradiation, restriction of tumor growth as well as enhanced resistance to subsequent infection with various microbial pathogens (Nowotny and Behling 1982). The first reports on the curative effect of application of bacterial products on infection stemmed from treatment of patients suffering from abdominal typhus with crude extracts of bacteria at the end of the 19th century (Rumpf 1893). By this time, the widespread use of fever therapy, i.e. the injection of pyrogenic bacterial preparations, for the treatment of various diseases was initiated. An excellent review on fever therapy was written by Nowotny (Nowotny and Behling 1982).

The use of animal models to study the mechanisms underlying enhancement of host defenses by bacterial products was initiated in 1892 by Kanthack, who reported pyrogen-induced changes on leukocytes after injection of *Vibrio metchnikorii* filtrates into rabbits (Kanthack 1892). In 1955, Rowley was the first to describe increased resistance of mice to bacterial infection after administration of *E. coli* cell wall extracts 48h prior to challenge (Rowley 1955). The same protection was afforded when isolated endotoxin was injected instead of cell walls (Rowley 1956). Subsequently, this phenomenon of reduced susceptibility after endotoxin application was extended to infections with other bacterial species and even some viral pathogens (Shilo 1959). Pretreatment with endotoxin or cellular components of Gram-negative bacteria induced nonspecific protection against infection with a number of different extra- and intracellular bacteria including both Gram-negative and Gram-positive species (Boehme and Dubos 1956; Dubos and Schaedler 1956; Dubos and Schaedler 1957; Parant 1968; Parant 1980; Parant et al. 1980; Rowley 1960), reviewed in detail by Shilo (Shilo 1959) (Table 4). More recently, increased resistance of LPS-pretreated animals to lethality and organ damage associated with multi-germ sepsis, induced e.g. by CLP was reported (Neviere et al. 1999; Salkowski et al. 1998; Urbaschek et al. 1984; Urbaschek and Urbaschek 1985). Experiments performed by Rayhane et al. corroborated the notion that increased resistance is nonspecific by demonstrating improved survival and decreased fungal burden of LPS pretreated mice with disseminated *Cryptococcus neoformans* infection (Rayhane et al. 2000).

Table 4. Effect of endotoxin on host defense

	Challenge	Survival	Mechanism	Reference
Gram-negative bacteria	*Escherichia coli* *Proteus vulgaris* *Pseudomonas aeruginosa* *Klebsiella pneumoniae* *Salmonella typhi*	↑ ↑ ↑ ↑ ↑	CFU↓ (*S. typhi*), Properdin levels ↑	(Landy and and Pillemer 1956)
	Escherichia coli	↑	serum bactericidal activity ↑, RES phagocytosis ↑	(Rowley 1956)
	Salmonella typhi	↑		(Tully et al. 1965)
	Salmonella dublin	↑	not transferred with serum, i.e. antibody independent	(Hill et al. 1974)
	Salmonella typhimurium *Salmonella enteritidis*	↑ ↑	CFU ↓, bacterial clearance↑	(Nakano et al. 1984)
	Salmonella typhimurium		serum opsonic activity ↑ MΦ phagocytic acitivity ↑	(Rowley 1960)
	Salmonella typhimurium	↑	CFU ↓, phagocyte accumulation ↑ phagocytosis/bactericidal activity of PMΦ ↑	(Onozuka et al. 1993)
	Salmonella typhimurium	↑	CFU ↓, PMN recruitment ↑, PMN accumulation ↑, RES phagocytosis ↑	(Lehner et al. 2001a)
	Klebsiella pneumoniae	↑	CFU ↓, RES phagocytosis ↑	(Galelli et al. 1977; Parant et al. 1976; Parant et al. 1980)

	Klebsiella pneumoniae	←		(Kiser et al. 1956)
	Pasteurella tularensis	←		(Pannell 1956a; Pannell 1956b)
Mycobacteria	Mycobacterium fortuitum	←	CFU ↓, RES phagocytosis↑	(Boehme and Dubos 1956)
Gram-positive bacteria	Staphylococcus aureus	←	CFU ↓	(Dubos and Schaedler 1956)
	Staphylococcus aureus Streptococcus pyogenes Diplococcus pneumoniae	no effect	porperdin insensitive bacteria	(Landy and and Pillemer 1956)
	Staphylococcus aureus	←	cytokine production ↓	(Astiz et al. 1994a)
	Streptococcus agalactiae	←	not transferred with serum → antibody independent	(Hill et al. 1974)
	Listeria monocytogenes	←	CFU ↓,	(Lehner et al. 2001a)
Mixed infection	CLP	←	CFU ↓, granulopoiesis ↑	(Urbaschek et al. 1984; Urbaschek and Urbaschek 1985)
	CLP	←	cytokine production ↓	(Astiz et al. 1994b)
Fungi	Cryptococcus neoformans	←	CFU ↓, TNF-mediated effect	(Rayhane et al. 2000)
Parasites	Plasmodium berghei	←		(MacGregor et al. 1969)

Mechanisms of Enhanced Host Defense

Humoral Factors

It has been demonstrated that enhanced resistance after LPS injection was associated with increased bactericidal activity of serum towards certain Gram-negative bacteria (Rowley 1956). Since evidence was provided that LPS administration enhanced serum bactericidal activity mainly towards properdin-sensitive organisms, increased serum properdin levels were suggested to be a major mechanism of LPS-induced resistance (Landy and Pillemer 1956). This view was questioned later by findings that LPS pretreatment afforded protection also to properdin-insensitive organisms such as Gram-positive bacteria. Moreover, alterations in host resistance against bacterial infection were not always paralleled by serum properdin levels (Howard et al. 1958). We recently provided further evidence against a major role of the complement system in LPS-induced increased resistance by demonstrating LPS-induced nonspecific resistance to *S. typhimurium* and *L. monocytogenes* in the absence of any changes in complement activity as determined in a sheep erythrocyte lysate assay (Lehner et al. 2001a). Furthermore, depletion of the central C3 protein of the complement cascade by administration of cobra venom factor did not abolish the protective effect of LPS pretreatment on *S. typhimurium* infection (Lehner et al. 2001a). However, increased serum opsonization activity after LPS administration was reported by several authors (Rowley 1960; Ruggiero et al. 1980). In sum, enhanced resistance to infection is associated with increased serum bactericidal or opsonization activity in some models, although direct proof for a critical contribution of the complement system is still lacking.

Macrophages

On the cellular level of host defense, LPS injection is associated with a transient depression of RES activity, followed by a longer lasting period of enhanced clearance of carbon particles, radioactive LPS, labelled chromium phosphate and viable or heat-killed bacteria by the RES (Galelli et al. 1977; Lehner et al. 2001a; Nakano et al. 1984). Detailed studies by Chedid's group demonstrated that irradiation- and cyclophosphamide-resistant cells mediated improved survival, enhanced RES phagocytic activity and reduced bacterial burden associated with LPS pretreatment of mice subsequently submitted to an otherwise lethal *Klebsiella pneumoniae* infection (Galelli et al. 1977; Parant et al. 1976; Parant et al. 1980). Although definite protection of irradiated mice by LPS injection additionally depended on a further, bone-marrow derived cell type not identical with T lymphocytes, their experiments strongly supported the notion that activation of RES macrophages was a major mechanism of LPS-induced host defense against *Klebsiella pneumoniae* (Galelli et al. 1977; Parant et al. 1976).

We have shown recently that enhanced hepatic phagocytosis of bacteria in LPS-pretreated mice was associated with increased numbers of Kupffer cells, the resident macrophage population of the liver (Lehner et al. 2001a). Increased Kupffer cell numbers were also reported for LPS-tolerant rats (Bautista and Spitzer 1995). Direct evidence for a contribution of Kupffer cells in LPS-stimulated clearance of bacteria was derived from experiments using chlodronate-liposomes to deplete liver macrophages prior to injection of bacteria (Lehner et al. 2001a). Ruggiero et al. used isolated perfused rat livers to demonstrate increased hepatic uptake of *Escherichia coli* after *in vivo* LPS pretreatment due to enhanced phagocytic activity of the liver and improved opsonization by the serum (Ruggiero et al. 1980). Besides an increase in Kupffer cell numbers, enhanced phagocytic activity of individual liver macrophages could account for improved hepatic clearance after LPS-treatment, as demonstrated by Hafenrichter et al. for isolated Kupffer cells from LPS-pretreated rats (Hafenrichter et al. 1994). Accordingly, peritoneal macrophages exposed to LPS *in vivo* or *in vitro* showed accelerated phagocytosis of *Salmonella typhimurium in vitro* (Rowley 1960). In contrast to studies using murine peritoneal macrophages where an enhancement of oxidative burst activity was reported (Leon et al. 1992; Onozuka et al. 1993), Kupffer cells from LPS-pretreated rats displayed decreased generation of superoxide anions (Bautista and Spitzer 1995). However, our unpublished data indicate improved antibacterial activity of Kupffer cells from endotoxin-tolerant mice.

Neutrophilic Granulocytes

LPS induces a plethora of chemokines leading to accumulation of leukocytes, consisting mainly of neutrophilic granulocytes, at the site of LPS administration. This is of importance when bacteria are injected at the site of previous LPS administration, since the microorganisms are confronted immediately with a large number of phagocytes absent in the naive host. We recently demonstrated that intraperitoneal accumulation of leukocytes and enhanced inactivation of intraperitoneally injected *Salmonella typhimurium* during the first hours post infection was strictly dependent upon the route of LPS pretreatment. Similar results were obtained by Astiz et al. who studied the therapeutic value of administration of monophosphoryl lipid A (MPL), a detoxified LPS derivative, to mice prior to induction of peritonitis by CLP. In their setting, intraperitoneal (*i.p.*) injections of MPL were more effective in reducing mortality than intravenous (*i.v.*) MPL administration (Astiz et al. 1994b). However, activation of resident peritoneal macrophages by *i.p.* LPS injection could also account for the improved antibacterial activity.

It has long been known that endotoxin is a potent stimulator of hematopoesis. Post-endotoxin serum was shown to have potent colony-stimulating factor (CSF) activity *in vitro* as well as *in vivo*, when transferred to naive animals (Butler et al. 1978; Butler and Nowotny 1976; Chang et al. 1974). Intensive studies on radioprotection by previous administration of endotoxin suggested an important role of accelerated hematopoiesis, as reviewed by Nowotny et al. (Nowotny and Behling 1982). Administration of LPS resulted in increased white blood cell numbers (He et al. 1992; Lehner et al. 2001a), neutrophilia (Kiani et al. 1997; Lehner et al. 2001a) and augmented numbers of monocyte/macrophage precursors in the bone marrow (Madonna et al. 1986; Madonna and Vogel 1985). We demonstrated that endotoxin-pretreated mice displayed elevated numbers of circulating neutrophils throughout the course of *Salmonella* infection, indicating improved recruitment from the bone marrow and/or decreased rate of apoptosis of these cells after LPS treatment (Lehner et al. 2001a). A critical role of diminished neutrophil apoptosis for the survival benefit associated with endotoxin pretreatment prior to induction of multi-germ peritonitis was suggested recently (Feterowski et al. 2001). This is in line with previous findings by Yamamoto, showing a delay of neutrophil apoptosis by LPS and LPS-induced cytokines *in vivo* and *in vitro* (Yamamoto et al. 1993).

Besides an increase in overall PMN numbers, enhanced anti-microbial activity of the individual PMN could contribute to enhanced immune defense of the LPS-pretreated host. Our unpublished data indicate an increased oxidative burst response of blood PMN from LPS-tolerant mice upon stimulation *ex vivo*. The view that neutrophils play a decisive role in LPS-induced resistance to infection is substantiated by our findings that PMN depletion partially abrogated the survival benefit of LPS-pretreated mice infected with *Salmonella typhimurium* (Lehner et al. 2001a).

Lymphocytes

Activation of lymphocytes by LPS or LPS-induced mediators is well documented, and Galelli et al. demonstrated that definite protection by LPS treatment of irradiated mice required bone-marrow derived radiosensitive cells (Galelli et al. 1977). However, the adaptive immune system seems to be of minor importance for the establishment of the early phase of LPS-induced nonspecific resistance as suggested by experiments performed with athymic or SCID-mice which showed protection in spite of lacking functional T-and B-lymphocytes (Galelli et al. 1977, our own results).

Mediators of Nonspecific Resistance

Many of the effects of endotoxin are mediated by endogenous mediators such as cytokines, arachidonic acid metabolites, reactive oxygen or nitrogen radicals. The role of autocrine mediators in the process of inducing or maintaining macrophage refractoriness is still under debate. Similarly, there is evidence that LPS-enhanced nonspecific resistance is the result of the biological activity of several cytokines produced in response to LPS injection.

Injection of IL-1 improved survival of mice infected subsequently with *Listeria monocytogenes* (Morikage et al. 1990), *Pseudomonas aeruginosa* (Morikage et al. 1990; Vogels et al. 1995; Vogels et al. 1994a; Vogels et al. 1994b; Vogels et al. 1992), *Klebsiella pneumoniae* (Morikage et al. 1990; Vogels et al. 1994b; Vogels et al. 1992), *Escherichia coli* (Lange et al. 1992) and in the sepsis model of CLP (Alexander et al. 1991b; O'Reilly et al. 1992a). Furthermore, the combination of IL-1 and TNF reduced mortality and bacterial load of mice infected with *E. coli* at 20-fold the LD50 (Cross et al. 1989). Pretreatment with IL-1, GM-CSF or G-CSF improved survival after aerosol pneumococcal challenge (Hebert and O'Reilly 1996; Hebert et al. 1997; Hebert et al. 1990). This effect could be due to enhanced microbicidal activity of alveolar macrophages and improved clearance of blood-borne pathogens of cytokine-pretreated mice (Hebert et al. 1994). The beneficial effect of G-CSF treatment prior to induction of bacterial peritonitis or *L. monocytogenes* infection was probably mediated via the recruitment or activation of PMN (Barsig et al. 1996; O'Reilly et al. 1992b; Serushago et al. 1992; Villa et al. 1998). Extensive studies on the beneficial effect of cytokine pretreatment on resolution of infection were performed in the model of *Salmonella typhimurium* infection of mice employed also in our studies. It has been shown that administration of TNF resulted in improved survival of otherwise lethal bacterial challenge (Nakano et al. 1990; Nauciel and Espinasse-Maes 1992). Protection against salmonella infection was also conferred by pretreatment with IFNγ (Nauciel and Espinasse-Maes 1992), IL-18 (Mastroeni et al. 1999) or TGFβ (Galdiero et al. 1999), IL-1 or a combination of IL-1 and TNF (Morrissey and Charrier 1994; Morrissey et al. 1995). Since LPS administration induces the formation of all of these mediators, it is feasible that nonspecific resistance is conferred via endogenous formation of these cytokines. However, direct proof for this hypothesis has not been provided yet. Studies on the role of LPS-induced cytokines in enhancing resis-

tance to infection are hampered by the fact that normal host defense initiated by the pathogen itself also depends on an intact cytokine response.
Thus, cytokine-deficient mice are unsuitable and the use of cytokine-specific
antibodies requires detailed titration experiments in order to selectively
neutralize only LPS-induced cytokines during the pretreatment phase but
not during infection. Furthermore, because of the plethora of cytokines with
similar protective effect, it is unlikely that neutralization of single mediators
will abrogate the beneficial effect of LPS-pretreatment.

Outlook

The finding that LPS-pretreated animals were protected against the toxicity
of endotoxin in models of septic shock and sepsis and displayed even enhanced resistance to bacterial infection, suggests the therapeutic use of endotoxin tolerance induction as a sepsis prophylaxis. However, the well-
known side-effects of endotoxin injection ranging from fever to potentially
fatal systemic inflammatory responses hamper the clinical use of endotoxin
administration. The use of detoxified derivatives of LPS such as synthetic
lipid A could avoid this risk. Several studies have demonstrated that these
substances retain the ability to protect against shock and bacterial infection
despite strongly decreased toxicity (Astiz et al. 1994a; Astiz et al. 1995; Astiz
et al. 1994b; Gustafson et al. 1995; Hamilton-Davies and Webb 1995; Henricson et al. 1990; Lam et al. 1991a; Lam et al. 1991b; Madonna et al. 1986; Rudbach et al. 1994; Salkowski et al. 1998; Schutze et al. 1994; Yao et al. 1994).
Future experiments will evaluate the clinical value of prophylactic induction
of LPS tolerance in reducing the risk of postoperative sepsis.

Abbreviations

AP-1	activation protein-1
CFU	colony forming units
CLP	cecal ligation and puncture
GalN	D-galactosamine
IFN	interferon
I-kB	inhibitor of NF–κB
IL	interleukin
i.p.	intraperitoneal
IL-1R	interleukin-1 receptor
IRAK	interleukin-1 receptor-associated kinase
i.v.	intravenous
LPS	lipopolysaccharide

LTA	lipoteichoic acid
MALP-2	macrophage-activating lipoprotein-2
MAP	mitogen-activated protein
MPL	monophosphoryl lipid A
mu	murine
MyD88	myeloid differentiation protein
NF–κB	nuclear factor kappa B
NIK	NF–κB-inducing kinase
PBMC	peripheral blood mononuclear cells
PBS	phosphate buffered saline
PGE$_2$	prostaglandin E$_2$
PMN	polymorphonuclear cells
r	recombinant
RES	reticuloendothelial system
S. aureus	Staphylococcus aureus
S. typhimurium	Salmonella typhimurium
TLR	Toll like receptor
TNF	tumor necrosis factor
TNF-R	tumor necrosis factor receptor
TRAF6	TNF receptor-activated factor 6

References

Alexander HR, Doherty GM, Buresh CM, Venzon DJ, Norton JA (1991a) A recombinant human receptor antagonist to interleukin 1 improves survival after lethal endotoxemia in mice. J Exp Med 173:1029-1032

Alexander HR, Doherty GM, Fraker DL, Block MI, Swedenborg JE, Norton JA (1991b) Human recombinant interleukin-1 alpha protection against the lethality of endotoxin and experimental sepsis in mice. J Surg Res 50:421-424

Ancuta P, Fahmi H, Pons JF, Le Blay K, Chaby R (1997) Involvement of the membrane form of tumour necrosis factor-alpha in lipopolysaccharide-induced priming of mouse peritoneal macrophages for enhanced nitric oxide response to lipopolysaccharide. Immunology 92:259-266

Annenkov AY, Baranova FS (1991) Lipopolysaccharide-dependent and Lipopolysaccharide-independent pathways of monocyte desensitisation to Lipopolysaccharides. J Leukocyte Biol 50:215-222

Astiz ME, Galera A, Saha DC, Carpati C, Rackow EC (1994a) Monophosphoryl lipid A protects against gram-positive sepsis and tumor necrosis factor. Shock 2:271-274

Astiz ME, Rackow EC, Still JG, Howell ST, Cato A, Von Eschen KB, Ulrich JT, Rudbach JA, McMahon G, Vargas R, et al. (1995) Pretreatment of normal humans with monophosphoryl lipid A induces tolerance to endotoxin: a prospective, double-blind, randomized, controlled trial. Crit Care Med 23:9-17

Astiz ME, Saha DC, Carpati CM, Rackow EC (1994b) Induction of endotoxin tolerance with monophosphoryl lipid A in peritonitis: importance of localized therapy. J Lab Clin Med 123:89-93

Atkins E (1960) Pathogenesis of fever. Physiol Rev 40:580-646

Baer M, Dillner A, Schwartz RC, Sedon C, Nedospasov S, Johnson PF (1998) Tumor necrosis factor alpha transcription in macrophages is attenuated by an autocrine factor that preferentially induces NF-kappaB p50. Mol Cell Biol 18:5678-5689

Balkhy HH, Heinzel FP (1999) Endotoxin fails to induce IFN-gamma in endotoxin-tolerant mice: deficiencies in both IL-12 heterodimer production and IL-12 responsiveness. J Immunol 162:3633-3638

Barsig J, Bundschuh DS, Hartung T, Bauhofer AM, Sauer A, and Wendel A (1996) Control of fecal peritoneal infection in mice by colony-stimulating factors. The Journal of Infectious Diseases 174:790-799

Basu S, Dunn AR, Marino MW, Savoia H, Hodgson G, Lieschke GJ, Cebon J (1997) Increased tolerance to endotoxin by granulocyte-macrophage colony-stimulating factor-deficient mice. J Immunol 159:1412-1417

Bautista AP, Spitzer JJ (1995) Acute endotoxin tolerance downregulates superoxide anion release by the perfused liver and isolated hepatic nonparenchymal cells. Hepatology 21:855-862

Baykal A, Kaynaroglu V, Hascelik G, Sayek I, Sanac Y (1999) Epinephrine and endotoxin tolerance differentially modulate serum cytokine levels to high-dose lipopolysaccharide challenge in a murine model. Surgery 125:403-410

Beeson PB (1946) Development of tolerance to typhoid bacterial pyrogen and its abolition by reticulo-endothelial blockade. Proc Soc Exp Biol Med 61:248-250

Beeson PB (1947) Tolerance to bacterial pyrogens. J Exp Med 86:29-44

Berg DJ, Kuhn R, Rajewsky K, Muller W, Menon S, Davidson N, Grunig G, Rennick D (1995) Interleukin-10 is a central regulator of the response to LPS in murine models of endotoxic shock and the Shwartzman reaction but not endotoxin tolerance. J Clin Invest 96:2339-2347

Beutler B, Krochin N, Milsark IW, Luedke C, Cerami A (1986) Control of cachectin (tumor necrosis factor) synthesis: mechanisms of endotoxin resistance. Science 232:977-980

Beutler B, Milsark IW, Cerami AC (1985) Passive immunization against cachectin/tumor necrosis factor protects mice from lethal effect of endotoxin. Science 229:869-871

Blackwell TS, Blackwell TR, Christman JW (1997) Impaired activation of nuclear factor-kappaB in endotoxin-tolerant rats is associated with down-regulation of chemokine gene expression and inhibition of neutrophilic lung inflammation. J Immunol 158:5934-5940

Boehme D, Dubos RJ (1956) The effect of bacterial constituents on the resistance of mice to heterologous infection and on the activity of their reticulo-endothelial system. J. Exp. Med. 107:523-536

Bogdan C, Vodovotz Y, Paik J, Xie QW, Nathan C (1993) Traces of bacterial lipopolysaccharide suppress IFN-gamma-induced nitric oxide synthase gene expression in primary mouse macrophages. J Immunol 151:301-309

Bohlinger I, Leist M, Barsig J, Uhlig S, Tiegs G, Wendel A (1995) Interleukin-1 and nitric oxide protect against tumor necrosis factor alpha-induced liver injury through distinct pathways. Hepatology 22:1829-1837

Bohuslav J, Kravchenko VV, Parry GC, Erlich JH, Gerondakis S, Mackman N, Ulevitch RJ (1998) Regulation of an essential innate immune response by the p50 subunit of NF-kappaB. J Clin Invest 102:1645-1652

Bone RC (1996) Sir Isaac Newton, sepsis, SIRS, and CARS. Crit Care Med 24:1125-1128

Bone RC, Grodzin CJ, Balk RA (1997) Sepsis: a new hypothesis for pathogenesis of the disease process. Chest 112:235-243

Bowling WM, Hafenrichter DG, Flye MW, Callery MP (1995) Endotoxin tolerance alters phospholipase C-gamma 1 and phosphatidylinositol-3'-kinase expression in peritoneal macrophages. J Surg Res 58:592-598

Bundschuh DS (1997) Immunomodulation in murine models of the systemic inflammatory response syndrome. Doktorarbeit an der Universität Konstanz.

Bundschuh DS, Barsig J, Hartung T, Randow F, Döcke WD, Volk HD, and Wendel A (1997) Granulocyte-Macrophage Colony-Stimulating Factor and IFN-gamma restore the systemic TNF-alpha response to endotoxin in lipopolysaccharide-desensitized mice. J Immunol 158:2862-2871

Burrell R (1994) Human responses to bacterial endotoxin. Circulatory Shock 43:137-153

Busam KJ, Schulze-Specking A, Decker K (1991) Endotoxin-refractory liver macrophages secrete tumor necrosis factor-alpha upon viral infection. Biol Chem Hoppe Seyler 372:157-162

Butler RC, Abdelnoor AM, Nowotny A (1978) Bone marrow colony-stimulating factor and tumor resistence-enhancing activity of postendotoxin mouse sera. Proc Natl Acad Sci 75:2893-2896

Butler RC, Nowotny A (1976) Colony stimulating factor (CSF)-containing serum has anti-tumor effects. IRCS Medical Science 4:206

Car BD, Eng VM, Schnyder B, Ozmen L, Huang S, Gallay P, Heumann D, Aguet M, Ryffel B (1994) Interferon gamma receptor deficient mice are resistant to endotoxic shock. J Exp Med 179:1437-1444

Cavaillon J-M (1995) The nonspecific nature of endotoxin tolerance. Trends Microbiol 3:320-324

Cavaillon JM, Munoz C, Marty C, Cabie A, Tamion F, Misset B, Carlet J, Fitting C (1993) Cytokine production by monocytes from patients with sepsis syndrome and by endotoxin-tolerant human monocytes. In: Levein J, Alving CR, Munford RS, Stütz PL (eds) Bacterial endotoxin: Recognition and effector mechanisms. Elsevier Science Publisher B.V., pp 275-286

Cavaillon J-M, Pitton C, Fitting C (1994) Endotoxin tolerance is not a LPS-specific phenomenon: partial mimicry with IL-1, IL-10, and TGFb. J Endotoxin Res 1:21-29

Centanni E (1894) Untersuchungen über das Infectionsfieber - das Fiebergift der Bacterien. Dtsch. Med. Wochenschr. 20:148-150

Centanni E (1942) Immunitätserscheinungen im experimentellen Fieber mit besonderer Berücksichtigung des pyrogenen Stoffes aus Typhusbakterien. Klin. Wochenschr. 21:664-669

Chamulitrat W, Wang JF, Spitzer JJ (1995) Electron paramagnetic resonance investigations of nitrosyl complex formation during endotoxin tolerance. Life Sci 57:387-395

Chang CC, McCormick CC, Lin AW, Dietert RR, Sung YJ (1996) Inhibition of nitric oxide synthase gene expression in vivo and in vitro by repeated doses of endotoxin. Am J Physiol 271:G539-548

128   M. D. Lehner and T. Hartung

Chang H, Thompson JJ, Nowotny A (1974) Release of colony stimulating factor (CSF) by non-endotoxic breakdown products of bacterial lipopolysaccharides. Immunol Commun 3:401-409

Chautard T, Spinedi E, Voirol M, Pralong FP, Gaillard RC (1999) Role of glucocorticoids in the response of the hypothalamo-corticotrope, immune and adipose systems to repeated endotoxin administration. Neuroendocrinology 69:360-369

Chedid L, Parant M, Boyer F, Skarnes RC (1964) Nonspecific host responses in tolerance to the lethal effect of endotoxins. In: Landy M, Braun W (eds) Bacterial Endotoxin. Rutgers University Press, New Brunswick, NJ, pp 500-516

Coelho MM, Luheshi G, Hopkins SJ, Pela IR, Rothwell NJ (1995) Multiple mechanisms mediate antipyretic action of glucocorticoids. Am J Physiol 269:R527-535

Coffee KA, Halushka PV, Ashton SH, Tempel GE, Wise WC, Cook JA (1992) Endotoxin tolerance is associated with altered GTP-binding protein function. J Appl Physiol 73:1008-1013

Cross A, Asher L, Seguin M, Yuan L, Kelly N, Hammack C, Sadoff J, Gemski P, Jr. (1995) The importance of a lipopolysaccharide-initiated, cytokine-mediated host defense mechanism in mice against extraintestinally invasive Escherichia coli. J Clin Invest 96:676-686

Cross AS, Sadoff JC, Kelly N, Bernton E, Gemski P (1989) Pretreatment with recombinant murine tumor necrosis factor alpha/cachectin and murine interleukin 1 alpha protects mice from lethal bacterial infection. J Exp Med 169:2021-2027

Dai WJ, Bartens W, Kohler G, Hufnagel M, Kopf M, Brombacher F (1997) Impaired macrophage listericidal and cytokine activities are responsible for the rapid death of Listeria monocytogenes-infected IFN- gamma receptor-deficient mice. J Immunol 158:5297-5304

de Waal Malefyt R, Abrams J, Bennett B, Figdor CG, de Vries JE (1991) Interleukin 10(IL-10) inhibits cytokine synthesis by human monocytes: an autoregulatory role of IL-10 produced by monocytes. J Exp Med 174:1209-1220

Dinarello CA (1997) Proinflammatory and anti-inflammatory cytokines as mediators in the pathogenesis of septic shock. Chest 112:321S-329S

Dinarello CA (2000) Proinflammatory cytokines. Chest 118:503-508

Dinarello CA, Bodel PT, Atkins E (1968) The role of the liver in the production of fever and in pyrogenic tolerance. Trans Assoc Am Physicians 81:334-344

Docke WD, Randow F, Syrbe U, Krausch D, Asadullah K, Reinke P, Volk HD, Kox W (1997) Monocyte deactivation in septic patients: restoration by IFN-gamma treatment. Nat Med 3:678-681

Dubos RJ, Schaedler RW (1956) Reversible changes in the susceptibility of mice to bacterial infections. J. Exp. Med. 104:53-65

Dubos RJ, Schaedler RW (1957) Effects of cellular constituents of mycobacteria on the resistance of mice to hererologous infections. J. Exp. Med. 106:703-717

Erroi A, Fantuzzi G, Mengozzi M, Sironi M, Orencole SF, Clark BD, Dinarello CA, Isetta A, Gnocchi P, Giovarelli M, Giovarelli M, Ghezzi P (1993) Differential regulation of cytokine production in lipopolysaccharide tolerance in mice. Infect Immun 61:4356-4359

Ertel W, Kremer J-P, Kenney J, Steckholzer U, Jarrar D, Trentz O, Schildberg FW (1995) Downregulation of proinflammatory cytokine release in whole blood from septic patients. Blood 85:1341-1347

Evans GF, Zuckerman SH (1991) Glucocorticoid-dependent and -independent mechanisms involved in lipopolysaccharide tolerance. Eur J Immunol 21:1973-1979

Faggioni R, Fantuzzi G, Villa P, Buurman W, van Tits LJ, Ghezzi P (1995) Independent down-regulation of central and peripheral tumor necrosis factor production as a result of lipopolysaccharide tolerance in mice. Infect Immun 63:1473-1477

Fahmi H, Chaby R (1993) Selective refractoriness of macrophages to endotoxin-induced production of tumor necrosis factor, elicited by an autocrine mechanism. J Leukoc Biol 53:45-52

Favorite GO, Morgan HR (1942) Effects produced by the intravenous injection in man of a toxic antigenic material derived from Eberthella thyphosa: clinical, hematological, chemical and serological studies. J. Clin. Invest. 21:589-599

Feterowski C, Weighardt H, Emmanuilidis K, Hartung T, Holzmann B (2001) Immune protection against septic peritonitis in endotoxin-primed mice is related to reduced neutrophil apoptosis. Eur. J. Immunol. in press

Flach R, Schade FU (1997) Peritoneal macrophages from endotoxin-tolerant mice produce an inhibitor of tumor necrosis factor alpha synthesis and protect against endotoxin shock. J. Endotoxin Res. 4:241-250

Flohe S, Dominguez Fernandez E, Ackermann M, Hirsch T, Borgermann J, Schade FU (1999) Endotoxin tolerance in rats: expression of TNF-alpha, IL-6, IL-10, VCAM-1 AND HSP 70 in lung and liver during endotoxin shock. Cytokine 11:796-804

Fraker DL, Stovroff MC, Merino MJ, Norton JA (1988) Tolerance to tumor necrosis factor in rats and the relationship to endotoxin tolerance and toxicity. J Exp Med 168:95-105

Frankenberger M, Pechumer H, Ziegler-Heitbrock HW (1995) Interleukin-10 is upregulated in LPS tolerance. J Inflamm 45:56-63

Freedman HH (1960a) Further studies on passive transfer of tolerance to pyrogenicity of bacterial endotoxin. J. Exp. Med. 112:619-634

Freedman HH (1960b) Passive transfer of tolerance to pyrogenicity of bacterial endotoxin. J. Exp. Med. 111:453-463

Freudenberg MA, Galanos C (1988) Induction of tolerance to lipopolysaccharide (LPS)-D-galactosamine lethality by pretreatment with LPS is mediated by macrophages. Infect Immun 56:1352-1357

Freudenberg MA, Keppler D, Galanos C (1986) Requirement for lipopolysaccharide-responsive macrophages in galactosamine-induced sensitization to endotoxin. Infect Immun 51:891-895

Gahring LC, Daynes RA (1986) Desensitization of animals to the inflammatory effects of ultraviolet radiation is mediated through mechanisms which are distinct from those responsible for endotoxin tolerance. J Immunol 136:2868-2874

Galanos C, Freudenberg MA (1993) Mechanisms of endotoxin shock and endotoxin hypersensitivity. Immunobiology 187:346-356

Galdiero M, Marcatili A, Cipollaro de l'Ero G, Nuzzo I, Bentivoglio C, Romano Carratelli C (1999) Effect of transforming growth factor beta on experimental Salmonella typhimurium infection in mice. Infect Immun 67:1432-1438

Galelli A, Parant M, Chedid L (1977) Role of radiosensitive and radioresistant cells in nonspecific resistance to infection of LPS-treated mice. J Reticuloendothel Soc 21:109-118

Geller P, Merril ER, Jawetz E (1954) Effects of cortisone and antibioticx on the lethal action of endotoxins in mice. Proc Soc Exp Biol Med 86:716-725

Goujon E, Parnet P, Laye S, Combe C, Dantzer R (1996) Adrenalectomy enhances pro-inflammatory cytokines gene expression, in the spleen, pituitary and brain of mice in response to lipopolysaccharide. Brain Res Mol Brain Res 36:53-62

Granowitz EV, Porat R, Mier JW, Orencole SF, Kaplanski G, Lynch EA, Ye K, Vannier E, Wolff SM, Dinarello CA (1993) Intravenous endotoxin suppresses the cytokine response of peripheral blood mononuclear cells of healthy humans. J Immunol 151:1637-1645

Greisman SE, Wagnr HN, Iio M, Hornick RB (1964) Mechanisms of endotoxin tolerance. II. Relationship between endotoxin and reticuloendothelial system phagocytic activity in man. J Exp Med 119:241-264

Greisman SE, Young EJ, Carozza FA, Jr. (1969) Mechanisms of endotoxin tolerance. V. Specificity of the early and late phases of pyrogenic tolerance. J Immunol 103:1223-1236

Greisman SE, Young EJ, Woodward WE (1966) Mechanisms of endotoxin tolerance. IV. Specificity of the pyrogenic refractory state during continuous intravenous infusions of endotoxin. J Exp Med 124:983-1000

Gustafson GL, Rhodes MJ, Hegel T (1995) Monophosphoryl lipid A as a prophylactic for sepsis and septic shock. Prog Clin Biol Res 392:567-579

Haas JG, Baeuerle PA, Riethmuller G, Ziegler-Heitbrock HW (1990) Molecular mechanisms in down-regulation of tumor necrosis factor expression. Proc Natl Acad Sci U S A 87:9563-9567

Hafenrichter DG, Roland CR, Mangino MJ, Flye MW (1994) The Kupffer cell in endotoxin tolerance: mechanisms of protection against lethal endotoxemia. Shock 2:251-256

Hagberg L, Hull R, Hull S, McGhee JR, Michalek SM, Svanborg Eden C (1984) Difference in susceptibility to gram-negative urinary tract infection between C3H/HeJ and C3H/HeN mice. Infect Immun 46:839-844

Hamilton-Davies C, Webb AR (1995) Monophosphoryl lipid A and endotoxin tolerance [letter]. Crit Care Med 23:1789-1790

Hartung T, Wendel A (1992) Endotoxin-inducible cytotoxicity in liver cell cultures--II. Demonstration of endotoxin-tolerance. Biochem Pharmacol 43:191-196

Haslberger A, Sayers T, Reiter H, Chung J, Schutze E (1988) Reduced release of TNF and PCA from macrophages of tolerant mice. Circ Shock 26:185-192

He W, Fong Y, Marano MA, Gershenwald JE, Yurt RW, Moldawer LL, Lowry SF (1992) Tolerance to endotoxin prevents mortality in infected thermal injury: association with attenuated cytokine response. J Infect Dis 165:859-864

Heagy W, Hansen C, Nieman K, Rodriguez JL, West MA (2000) Impaired mitogen-activated protein kinase activation and altered cytokine secretion in endotoxin-tolerant human monocytes. J Trauma 49:806-814

Hebert JC, O'Reilly M (1996) Granulocyte-macrophage colony-stimulating factor (GM-CSF) enhances pulmonary defenses against pneumococcal infections after splenectomy. J Trauma 41:663-666

Hebert JC, O'Reilly M, Barry B, Shatney L, Sartorelli K (1997) Effects of exogenous cytokines on intravascular clearance of bacteria in normal and splenectomized mice. J Trauma 43:875-879

Hebert JC, O'Reilly M, Gamelli RL (1990) Protective effect of recombinant human granulocyte colony-stimulating factor against pneumococcal infections in splenectomized mice. Arch Surg 125:1075-1078

Hebert JC, O'Reilly M, Yuenger K, Shatney L, Yoder DW, Barry B (1994) Augmentation of alveolar macrophage phagocytic activity by granulocyte colony stimulating factor and interleukin-1: influence of splenectomy. J Trauma 37:909-912

Henricson BE, Benjamin WR, Vogel SN (1990) Differential cytokine induction by doses of lipopolysaccharide and monophosphoryl lipid A that result in equivalent early endotoxin tolerance. Infect Immun 58:2429-2437

Henricson BE, Neta R, Vogel SN (1991) An interleukin-1 receptor antagonist blocks lipopolysaccharide-induced colony-stimulating factor production and early endotoxin tolerance. Infect Immun 59:1188-1191

Hershman MJ, Cheadle WG, Wellhausen SR, Davidson PF, Polk HC, Jr. (1990) Monocyte HLA-DR antigen expression characterizes clinical outcome in the trauma patient. Br J Surg 77:204-207

Hill AW, Hibbitt KG, Shears A (1974) Endotoxin tolerance and the induction of nonspecific resistance to Salmonella dublin and Streptococcus agalactiae infections in mice. Br J Exp Pathol 55:269-274

Hirohashi N, Morrison DC (1996) Low-dose lipopolysaccharide (LPS) pretreatment of mouse macrophages modulates LPS-dependent interleukin-6 production in vitro. Infect Immun 64:1011-1015

Hoshino K, Takeuchi O, Kawai T, Sanjo H, Ogawa T, Takeda Y, Takeda K, Akira S (1999) Cutting edge: Toll-like receptor 4 (TLR4)-deficient mice are hyporesponsive to lipopolysaccharide: evidence for TLR4 as the Lps gene product. J Immunol 162:3749-3752

Howard JG, Rowley D, and Wardlay AC (1958) Immunology 1:181-203

Howard M, Muchamuel T, Andrade S, Menon S (1993) Interleukin 10 protects mice from lethal endotoxemia. J Exp Med 177:1205-1208

Irikura VM, Hirsch E, Hirsh D (1999) Effects of interleukin-1 receptor antagonist overexpression on infection by Listeria monocytogenes. Infect Immun 67:1901-1909

Johnston CA, Greisman SE (1985) Mechanism of endotoxin tolerance. In: Hinshaw IB (ed). Elsevier Science Publishers, pp 359 bis 401 (Handbook of Endotoxin

Kagaya K, Watanabe K, Fukazawa Y (1989) Capacity of recombinant gamma interferon to activate macrophages for Salmonella-killing activity. Infect Immun 57:609-615

Kanthack AA (1892) Acute leukocytosis produced by bacterial products. Br. Med. J. 1:1301-1303

Kaplan E, Dinarello CA, Wakabayashi G, Burke JF, Connolly RS, Gelfand JA (1993) Interleukin-1 pretreatment protects against endotoxin-induced hypotension in rabbits: association with decreased tumor necrosis factor levels. J Infect Dis 167:244-247

Karp CL, Wysocka M, Ma X, Marovich M, Factor RE, Nutman T, Armant M, Wahl L, Cuomo P, Trinchieri G (1998) Potent suppression of IL-12 production from monocytes and dendritic cells during endotoxin tolerance. Eur J Immunol 28:3128-3136

Kastenbauer S, Ziegler-Heitbrock HW (1999) NF-kappaB1 (p50) is upregulated in lipopolysaccharide tolerance and can block tumor necrosis factor gene expression. Infect Immun 67:1553-1559

Kiani A, Tschiersch A, Gaboriau E, Otto F, Seiz A, Knopf HP, Stutz P, Farber L, Haus U, Galanos C, Mertelsmann R, Engelhardt R (1997) Downregulation of the proinflammatory cytokine response to endotoxin by pretreatment with the nontoxic lipid A analog SDZ MRL 953 in cancer patients. Blood 90:1673-1683

Kimmings AN, Pajkrt D, Zaajer K, Moojen TM, Meenan JK, ten Cate JW, van Deventer SJH (1996) Factors involved in early in vitro endotoxin hyporesponsiveness in human endotoxemia. J Endotoxin Res 3:283-289

Kiser JS, Lindh H, de Mello GC (1956) The effect of various substances on resistance to experimental infections. Ann NY Acad Sci 66:312-327

Klosterhalfen B, Horstmann-Jungemann K, Vogel P, Flohe S, Offner F, Kirkpatrick CJ, Heinrich PC (1992) Time course of various inflammatory mediators during recurrent endotoxemia. Biochem Pharmacol 43:2103-2109

Knopf HP, Otto F, Engelhardt R, Freudenberg MA, Galanos C, Herrmann F, Schumann RR (1994) Discordant adaptation of human peritoneal macrophages to stimulation by lipopolysaccharide and the synthetic lipid A analogue SDZ MRL 953. Down-regulation of TNF-alpha and IL-6 is paralleled by an up-regulation of IL-1 beta and granulocyte colony-stimulating factor expression. J Immunol 153:287-299

Kox WJ, Bone RC, Krausch D, Docke WD, Kox SN, Wauer H, Egerer K, Querner S, Asadullah K, von Baehr R, Volk HD (1997) Interferon gamma-1b in the treatment of compensatory anti-inflammatory response syndrome. A new approach: proof of principle. Arch Intern Med 157:389-393

Kraatz J, Clair L, Bellingham J, Wahlstrom K, Rodriguez JL, West MA (1998) Lipopolysaccharide pretreatment produces macrophage endotoxin tolerance via a serum-independent pathway. J Trauma 45:684-691

Kraatz J, Clair L, Rodriguez JL, West MA (1999a) *In vitro* macrophage endotoxin tolerance: defective *in vitro* macrophage map kinase signal transduction after LPS pretreatment is not present in macrophages from C3H/HeJ endotoxin resistant mice. Shock 11:58-63

Kraatz J, Clair L, Rodriguez JL, West MA (1999b) Macrophage TNF secretion in endotoxin tolerance: role of SAPK, p38, and MAPK. J Surg Res 83:158-164

Kreutz M, Ackermann U, Hauschildt S, Krause SW, Riedel D, Bessler W, Andreesen R (1997) A comparative analysis of cytokine production and tolerance induction by bacterial lipopeptides, lipopolysaccharides and Staphyloccous aureus in human monocytes. Immunology 92:396-401

Kunkel SL, Spengler M, May MA, Spengler R, Larrick J, Remick D (1988) Prostaglandin E2 regulates macrophage-derived tumor necrosis factor gene expression. J Biol Chem 263:5380-5384

Labeta MO, Durieux JJ, Spagnoli G, Fernandez N, Wijdenes J, Herrmann R (1993) CD14 and tolerance to lipopolysaccharide: biochemical and functional analysis. Immunology 80:415-423

Lam C, Schutze E, Hildebrandt J, Aschauer H, Liehl E, Macher I, Stutz P (1991a) SDZ MRL 953, a novel immunostimulatory monosaccharidic lipid A analog with an improved therapeutic window in experimental sepsis. Antimicrob Agents Chemother 35:500-505.

Lam C, Schutze E, Liehl E, Stutz P (1991b) Effect of SDZ MRL 953 on the survival of mice with advanced sepsis that cannot be cured by antibiotics alone. Antimicrob Agents Chemother 35:506-511.

Landy M, and Pillemer L (1956) Increased resistance to infection and accompanying alteration in properdin levels following administration of bacterial lipopolysaccharides. JExpMed 104:383-409

Lange JR, Alexander HR, Merino MJ, Doherty GM, Norton JA (1992) Interleukin-1 alpha prevention of the lethality of *Escherichia coli* peritonitis. J Surg Res 52:555-559

Langermans JAM, van Furth R (1994) Cytokines and the host defense against Listeria monocytogenes and Salmonella typhimurium. Biotherapy 7:169-178

Larsen NE, Sullivan R (1984) Interaction between endotoxin and human monocytes: characteristics of the binding of 3H-labeled lipopolysaccharide and 51Cr-labeled lipid A before and after the induction of endotoxin tolerance. Proc Natl Acad Sci U S A 81:3491-3495

LaRue KE, McCall CE (1994) A labile transcriptional repressor modulates endotoxin tolerance. J Exp Med 180:2269-2275

Learn CA, Mizel SB, McCall CE (2000) mRNA and protein stability regulate the differential expression of pro- and anti-inflammatory genes in endotoxin-tolerant THP-1 cells. J Biol Chem 275:12185-12193

Lehner MD, Ittner J, Bundschuh DS, van Rooijen N, Wendel A, Hartung T (2001a) Improved innate immunity of endotoxin-tolerant mice increases resistance to salmonella enterica serovar typhimurium infection despite attenuated cytokine response. Infect Immun 69:463-471

Lehner MD, Morath S, Michelsen KS, Schumann RR, Hartung T (2001b) Induction of cross-tolerance by lipopolysaccharide and highly purified lipoteichoic acid via different toll-like receptors independent of paracrine mediators. J Immunol 166:5161-5167.

Leon P, Redmond HP, Shou J, Daly JM (1992) Interleukin 1 and relationship to endotoxin tolerance. Arch Surg 127:146-151

Lepe-Zuniga JL, Klostergaard J (1990) Tolerance to endotoxin in vitro: independent regulation of interleukin- 1, tumor necrosis factor and interferon alpha production during in vitro differentiation of human monocytes. Lymphokine Res 9:309-319

Li L, Cousart S, Hu J, McCall CE (2000) Characterization of interleukin-1 receptor-associated kinase in normal and endotoxin-tolerant cells. J Biol Chem 275:23340-23345

Li MH, Seatter SC, Manthei R, Bubrick M, West MA (1994) Macrophage endotoxin tolerance: effect of TNF or endotoxin pretreatment. J Surg Res 57:85-92

Libert C, Van Bladel S, Brouckaert P, Shaw A, Fiers W (1991) Involvement of the liver, but not of IL-6, in IL-1-induced desensitization to the lethal effects of tumor necrosis factor. J Immunol 146:2625-2632

Libert C, Van Molle W, Brouckaert P, Fiers W (1996) alpha1-Antitrypsin inhibits the lethal response to TNF in mice. J Immunol 157:5126-5129

MacGregor RR, Sheagren JN, Wolff SM (1969) Endotoxin-induced modification of Plasmodium berghei infection in mice. J Immunol 102:131-139

Mackensen A, Galanos C, Engelhardt R (1991) Modulating activity of interferon-gamma on endotoxin-induced cytokine production in cancer patients. Blood 78:3254-3258

Mackensen A, Galanos C, Wehr U, Engelhardt R (1992) Endotoxin tolerance: regulation of cytokine production and cellular changes in response to endotoxin application in cancer patients. Eur Cytokine Netw 3:571-579

Madonna GS, Peterson JE, Ribi EE, Vogel SN (1986) Early-phase endotoxin tolerance: induction by a detoxified lipid A derivative, monophosphoryl lipid A. Infect Immun 52:6-11

Madonna GS, Vogel SN (1985) Early endotoxin tolerance is associated with alterations in bone marrow- derived macrophage precursor pools. J Immunol 135:3763-3771

Madonna GS, Vogel SN (1986) Induction of early-phase endotoxin tolerance in athymic (nude) mice, B-cell-deficient (xid) mice, and splenectomized mice. Infect Immun 53:707-710

Makhlouf M, Zingarelli B, Halushka PV, Cook JA (1998a) Endotoxin tolerance alters macrophage membrane regulatory G proteins. Prog Clin Biol Res 397:217-226

Makhlouf MA, Fernando LP, Gettys TW, Halushka PV, Cook JA (1998b) Increased prostacyclin and PGE2 stimulated cAMP production by macrophages from endo-toxin-tolerant rats. Am J Physiol 274:C1238-1244

Marik PE, Varon J (1998) The hemodynamic derangements in sepsis: implications for treatment strategies. Chest 114:854-860

Mastroeni P, Clare S, Khan S, Harrison JA, Hormaeche CE, Okamura H, Kurimoto M, Dougan G (1999) Interleukin 18 contributes to host resistance and gamma interferon production in mice infected with virulent *Salmonella typhimurium*. Infect Immun 67:478-483

Mathison JC, Virca GD, Wolfson E, Tobias PS, Glaser K, Ulevitch RJ (1990) Adapta-tion to bacterial lipopolysaccharide controls lipopolysaccharide-induced tumor necrosis factor production in rabbit macrophages. J Clin Invest 85:1108-1118

Mathison JC, Wolfson E, Ulevitch RJ (1988) Participation of tumor necrosis factor in the mediation of gram negative bacterial lipopolysaccharide-induced injury in rabbits. J Clin Invest 81:1925-1937

Matic M, Simon SR (1991) Tumor necrosis factor release from lipopolysaccharide-stimulated human monocytes: lipopolysaccharide tolerance in vivo. Cytokine 3:576-583

Matic M, Simon SR (1992) Effects of gamma interferon on release of tumor necrosis factor alpha from lipopolysaccharide-tolerant human monocyte-derived macro-phages. Infect Immun 60:3756-3762

Matsuura K, Ishida T, Setoguchi M, Higuchi Y, Akizuki S, Yamamoto S (1994a) Upregulation of mouse CD14 expression in Kupffer cells by lipopolysaccharide. J Exp Med 179:1671-1676

Matsuura M, Kiso M, Hasegawa A, Nakano M (1994b) Multistep regulation mecha-nisms for tolerance induction to lipopolysaccharide lethality in the tumor-necrosis-factor-alpha- mediated pathway. Application of non-toxic monosac-charide lipid A analogues for elucidation of mechanisms. Eur J Biochem 221:335-341

McCall CE, Grosso-Wilmoth LM, LaRue K, Guzman RN, Cousart SL (1993) Toler-ance to endotoxin-induced expression of the interleukin-1 beta gene in blood neutrophils of humans with the sepsis syndrome. J Clin Invest 91:853-861

Medvedev AE, Kopydlowski KM, Vogel SN (2000) Inhibition of lipopolysaccharide-induced signal transduction in endotoxin-tolerized mouse macrophages: dys-regulation of cytokine, chemokine, and toll-like receptor 2 and 4 gene expres-sion. J Immunol 164:5564-5574

Medzhitov R, Preston-Hurlburt P, Kopp E, Stadlen A, Chen C, Ghosh S, Janeway CA, Jr. (1998) MyD88 is an adaptor protein in the hToll/IL-1 receptor family signaling pathways. Mol Cell 2:253-258

Mengozzi M, Fantuzzi G, Sironi M, Bianchi M, Fratelli M, Peri G, Bernasconi S, Ghezzi P (1993) Early down regulation of TNF production by LPS tolerance in human monocytes: comparison with IL-1, IL-6, and IL-8. Lympho Cyt Res 12:231-236

Mengozzi M, Sironi M, Gadina M, Ghezzi P (1991) Reversal of defective IL-6 pro-duction in lipopolysaccharide-tolerant mice by phorbol myristate acetate. J Im-munol 147:899-902

Moore JN, Cook JA, Morris DD, Halushka PV, Wise WC (1990) Endotoxin-induced procoagulant activity, eicosanoid synthesis, and tumor necrosis factor production by rat peritoneal macrophages: effect of endotoxin tolerance and glucan. Circ Shock 31:281-295

Moreau SC, Skarnes RC (1973) Host resistence to bacterial endotoxemia: mechanisms in endotoxin-tolerant animals. J Infect Dis 128:122-133

Morikage T, Mizushima Y, Sakamoto K, Yano S (1990) Prevention of fatal infections by recombinant human interleukin 1 alpha in normal and anticancer drug-treated mice. Cancer Res 50:2099-2104

Morrissey PJ, Charrier K (1994) Treatment of mice with IL-1 before infection increases resistence to a lethal challenge with *salmonella typhimurium*. J Immunol 153:212-219

Morrissey PJ, Charrier K, Vogel, S.N. (1995) Exogenous tumor necrosis factor alpha and interleukin-1a increase resistance to salmonella typhimurium: efficacy is influenced by the ity and lps loci. Infect Immun 63:3196-3198

Morrow LE, McClellan JL, Conn CA, Kluger MJ (1993) Glucocorticoids alter fever and IL-6 responses to psychological stress and to lipopolysaccharide. Am J Physiol 264:R1010-1016

Munoz C, Carlet J, Fitting C, Misset B, Bleriot J-P, Cavaillon J-M (1991) Dysregulation of in vitro cytokine response by monocytes during sepsis. JClin Invest 88:1747-1754

Nakano M, Onozuka K, Saito-Taki T (1984) LPS-induced non-specific resistance to immunodefective CBA/N mice against Salmonella infection Immunopharmacology of Endotoxicosis. Walther de Gruyter and Co., Berlin, pp 115-132

Nakano M, Onozuka K, Yamasu H, Zhong WF, Nakano Y (1992) Protective effects of cytokines in murine Salmonellosis. In: Friedman Hea (ed) Microbial Infections. Plenum Press, New York, pp 89-95 (Microbial. Infections

Nakano Y, Onozuka K, Terada Y, Shinomiya H, and Nakano M (1990) Protective effect of recombinant tumor necrosis factor alpha in murine salmonellosis. J Immunol 144:1935-1941

Nauciel C, Espinasse-Maes F (1992) Role of gamma interferon and tumor necrosis factor alpha in resistance to salmonella typhimurium infection. Infect Immun 60:450-454

Neviere RR, Cepinskas G, Madorin WS, Hoque N, Karmazyn M, Sibbald WJ, Kvietys PR (1999) LPS pretreatment ameliorates peritonitis-induced myocardial inflammation and dysfunction: role of myocytes. Am J Physiol 277:H885-892

Nomura F, Akashi S, Sakao Y, Sato S, Kawai T, Matsumoto M, Nakanishi K, Kimoto M, Miyake K, Takeda K, Akira S (2000) Cutting edge: endotoxin tolerance in mouse peritoneal macrophages correlates with down-regulation of surface toll-like receptor 4 expression. J Immunol 164:3476-3479

Nowotny A, Behling UH (1982) Studies on host defenses enhanced by endotoxins: a brief review. Klinische Wochenschrift 60:735-739

O'Brien DP, Briles DE, Szalai AJ, Tu AH, Sanz I, Nahm MH (1999) Tumor necrosis factor alpha receptor I is important for survival from *Streptococcus pneumoniae* infections. Infect Immun 67:595-601

Ogle CK, Guo X, Chance WT, Ogle JD (1997) Induction of endotoxin tolerance in rat bone marrow cells by *in vivo* infusion of tumor necrosis factor. Crit Care Med 25:827-833

Onozuka K, Shimada S, Yamasu H, Osada Y, Nakano M (1993) Non-specific resistance induced by a low-toxic lipid A analogue, DT-5461, in murine salmonellosis. Int J Immunopharmacol 15:657-664

O'Reilly M, Silver GM, Davis JH, Gamelli RL, Hebert JC (1992a) Interleukin 1 beta improves survival following cecal ligation and puncture. J Surg Res 52:518-522

O'Reilly M, Silver GM, Greenhalgh DG, Gamelli RL, Davis JH, Hebert JC (1992b) Treatment of intra-abdominal infection with granulocyte colony- stimulating factor. J Trauma 33:679-682

Ozmen Le (1994) Interleukin 12, interferon gamma, and tumor necrosis factor alpha are the key cytokines of the generalized Shwartzman reaction. J Exp Med 180:907-915

Pannell L (1956a) Studies on protection against experimental tularemia in mice I. J Infect Dis 102:162-166

Pannell L (1956b) Studies on protection against experimental tularemia in mice II. J Infect Dis 102:167-173

Parant M (1968) [Study of opsonins in mice after endotoxin augmentation of their resistance to infection]. Ann Inst Pasteur (Paris) 115:264-278

Parant M (1980) Antimicrobial resistance enhancing activity of tumor necrosis serum factor induced by endotoxin in BCG-treated mice. Recent Results Cancer Res 75:213-219

Parant M, Chedid L (1971) [Sensitization of mice to endotoxins by adrenalectomy or by administration of actinomycin D]. C R Acad Sci Hebd Seances Acad Sci D 272:1308-1311

Parant M, Galelli A, Parant F, Chedid L (1976) Role of B-lymphocytes in nonspecific resistance to Klebsiella pneumoniae infection of endotoxin-treated mice. J Infect Dis 134:531-539

Parant MA, Parant FJ, Chedid LA (1980) Enhancement of resistance to infections by endotoxin-induced serum factor from Mycobacterium bovis BCG-infected mice. Infect Immun 28:654-659

Pfeffer K, Matsuyama T, Kundig TM, Wakeham A, Kishihara K, Shahinian A, Wiegmann K, Ohashi PS, Kronke M, Mak TW (1993) Mice deficient for the 55 kd tumor necrosis factor receptor are resistant to endotoxic shock, yet succumb to L. monocytogenes infection. Cell 73:457-467

Poltorak A, He X, Smirnova I, Liu MY, Huffel CV, Du X, Birdwell D, Alejos E, Silva M, Galanos C, Freudenberg M, Ricciardi-Castagnoli P, Layton B, Beutler B (1998) Defective LPS signaling in C3H/HeJ and C57BL/10ScCr mice: mutations in Tlr4 gene. Science 282:2085-2088

Quesenberry P, Halperin J, Ryan M, Stohlman F (1975) Tolerance to the granulocyte-releasing and colony-stimulating factor elevating effects of endotoxin. Blood 45:789-800

Qureshi ST, Lariviere L, Leveque G, Clermont S, Moore KJ, Gros P, Malo D (1999) Endotoxin-tolerant mice have mutations in Toll-like receptor 4 (Tlr4). J Exp Med 189:615-625

Randow F, Döcke WD, Bundschuh DS, Hartung T, Wendel A, and Volk HD (1997) In vitro prevention and reversal of lipopolysaccharide desensitization by IFN gamma, IL-12, and Granulocyte-Macrophage Colony Stimulating Factor. J Immunol 158:2911-2918

Randow F, Syrbe U, Meisel C, Krausch D, Zuckermann H, Platzer C, Volk HD (1995) Mechanism of endotoxin desensitization: involvement of interleukin 10 and transforming growth factor beta. J Exp Med 181:1887-1892

Rayhane N, Fitting C, Lortholary O, Dromer F, Cavaillon JM (2000) Administration of endotoxin associated with lipopolysaccharide tolerance protects mice against fungal infection. Infect Immun 68:3748-3753

Renz H, Gong JH, Schmidt A, Nain M, Gemsa D (1988) Release of tumor necrosis factor-alpha from macrophages. Enhancement and suppression are dose-dependently regulated by prostaglandin E2 and cyclic nucleotides. J Immunol 141:2388-2393

Riedel DD, Kaufmann SH (2000) Differential tolerance induction by lipoarabino-mannan and lipopolysaccharide in human macrophages. Microbes Infect 2:463-471

Rodrick ML, Moss NM, Grbic JT, Revhaug A, O'Dwyer ST, Michie HR, Gough DB, Dubravec D, Manson JM, Saporoschetz IB, et al. (1992) Effects of *in vivo* endotoxin infusions on *in vitro* cellular immun responses in humans. J Clin Immunol 12:440-450

Rogers TS, Halushka PV, Wise WC, Cook JA (1986) Differential alteration of lipoxygenase and cyclooxygenase metabolism by rat peritoneal macrophages induced by endotoxin tolerance. Prostaglandins 31:639-650

Rogers TS, Halushka PV, Wise WC, Cook JA (1989) Arachidonic acid turnover in peritoneal macrophages is altered in endotoxin-tolerant rats. Biochim Biophys Acta 1001:169-175

Roth J, Aslan T, Storr B, Zeisberger E (1997) Lack of cross tolerance between LPS and muramyl dipeptide in induction of circulating TNF-alpha and IL-6 in guinea pigs. Am J Physiol 273:R1529-1533

Roth J, Zeisberger E (1995) Endotoxin tolerance alters thermal response of guinea pigs to systemic infusions of tumor necrosis factor-alpha. Am. J. Physiol. 268:514-519

Rothe J, Lesslauer W, Lotscher H, Lang Y, Koebel P, Kontgen F, Althage A, Zinkernagel R, Steinmetz M, Bluethmann H (1993) Mice lacking the tumour necrosis factor receptor 1 are resistant to TNF- mediated toxicity but highly susceptible to infection by *Listeria monocytogenes*. Nature 364:798-802

Rowley D (1955) Stimulation of natural immunity to Escherichia coli infections. The Lancet 1:232

Rowley D (1956) Rapidly induced changes in the level of nonspecific immunity in laboratory animals. Brit. J. Exp. Path. 37:233

Rowley D (1960) The role of opsonins in non-specific immunity. J. Exp. Med. 111:137-144

Rudbach JA, Myers KR, Rechtman DJ, Ulrich JT (1994) Prophylactic use of monophosphoryl lipid A in patients at risk for sepsis. Prog Clin Biol Res 388:107-124

Ruggiero G, Andreana A, Utili R, Galante D (1980) Enhanced phagocytosis and bactericidal activity of hepatic reticuloendothelial system during endotoxin tolerance. Infect Immun 27:798-803

Rumpf T (1893) Die Behandlung des Typhus ab dominalis mit abgetödteten Culturen des Bacillus pyocyaneus. Dtsch. Med. Wchnschr. 19:987-989

Salkowski CA, Detore G, Franks A, Falk MC, Vogel SN (1998) Pulmonary and hepatic gene expression following cecal ligation and puncture: monophosphoryl lipid A prophylaxis attenuates sepsis-induced cytokine and chemokine expression and neutrophil infiltration. Infect Immun 66:3569-3578

Sanchez-Cantu L, Rode HN, Christou NV (1989) Endotoxin tolerance is associated with reduced secretion of tumor necrosis factor. Arch Surg 124:1432-1435; discussion 1435-1436

Sato S, Nomura F, Kawai T, Takeuchi O, Muhlradt PF, Takeda K, Akira S (2000) Synergy and cross-tolerance between toll-like receptor (TLR) 2- and TLR4-mediated signaling pathways. J Immunol 165:7096-7101

Scales WE, Chensue SW, Otterness I, Kunkel SL (1989) Regulation of monokine gene expression: prostaglandin E2 suppresses tumor necrosis factor but not inter-leukin-1 alpha or beta-mRNA and cell-associated bioactivity. J Leukoc Biol 45:416-421

Schade F, Schlegel J, Hofmann K, Brade H, Flach R (1996) Endotoxin-tolerant mice produce an inhibitor of tumor necrosis factor synthesis. J. Endotoxin Res. 3:455-462

Schutze E, Hildebrandt J, Liehl E, Lam C (1994) Protection of mice from mortality caused by living and heat-killed bacteria by SDZ MRL 953. Circ Shock 42:121-127.

Seatter SC, Li MH, Bubrick MP, West MA (1995) Endotoxin pretreatment of human monocytes alters subsequent endotoxin- triggered release of inflammatory mediators. Shock 3:252-258

Serushago BA, Yoshikai Y, Handa T, Mitsuyama M, Muramori K, Nomoto K (1992) Effect of recombinant human granulocyte colony-stimulating factor (rh G-CSF) on murine resistence against Listeria monocytogenes. Immunol 75:475-480

Severn A, Xu D, Doyle J, Leal LM, O'Donnell CA, Brett SJ, Moss DW, Liew FY (1993) Pre-exposure of murine macrophages to lipopolysaccharide inhibits the induc-tion of nitric oxide synthase and reduces leishmanicidal activity. Eur J Immunol 23:1711-1714

Shilo M (1959) Nonspecific resistance to infections. Ann Rev Microbiol 13:255-278

Shimauchi H, Ogawa T, Okuda K, Kusumoto Y, Okada H (1999) Autoregulatory effect of interleukin-10 on proinflammatory cytokine production by *Porphyro-monas gingivalis* lipopolysaccharide-tolerant human monocytes. Infect Immun 67:2153-2159

Smith WB, Noack L, Khew-Goodall Y, Isenmann S, Vadas MA, Gamble JR (1996) Transforming growth factor-beta 1 inhibits the production of IL-8 and the transmigration of neutrophils through activated endothelium. J Immunol 157:360-368

Steer JH, Kroeger KM, Abraham LJ, Joyce DA (2000) Glucocorticoids suppress tumor necrosis factor-alpha expression by human monocytic THP-1 cells by suppress-ing transactivation through adjacent NF-kappa B and c-Jun-activating transcrip-tion factor-2 binding sites in the promoter. J Biol Chem 275:18432-18440

Szabo C, Thiemermann C, Wu CC, Perretti M, Vane JR (1994) Attenuation of the induction of nitric oxide synthase by endogenous glucocorticoids accounts for endotoxin tolerance in vivo. Proc Natl Acad Sci U S A 91:271-275

Takano M, Yokoyama K, Yayama K, Itoh N, Okamoto H (1993) Endotoxin-induced enhancement of angiotensinogen synthesis in the liver: decreased response fol-lowing repeated endotoxin exposure. Biol Pharm Bull 16:917-920

Takasuka N, Matsuura K, Yamamoto S, Akagawa KS (1995) Suppression of TNF-alpha mRNA expression in LPS-primed macrophages occurs at the level of nu-clear factor-kappa B activation, but not at the level of protein kinase C or CD14 expression. J Immunol 154:4803-4812

Takasuka N, Tokunaga T, Akagawa KS (1991) Preexposure of macrophages to low doses of lipopolysaccharide inhibits the expression of tumor necrosis factor-alpha mRNA but not of IL-1 beta mRNA. J Immunol 146:3824-3830

Tiegs G, Barsig J, Matiba B, Uhlig S, Wendel A (1994) Potentiation by granulocyte macrophage colony-stimulating factor of lipopolysaccharide toxicity in mice. J Clin Invest 93:2616-2622

Tite JP, Dougan G, Chatfield SN (1991) The involvement of tumor necrosis factor in immunity to Salmonella infection. J Immunol 147:3161-3164

Tominaga K, Saito S, Matsuura M, Nakano M (1999) Lipopolysaccharide tolerance in murine peritoneal macrophages induces downregulation of the lipopolysaccharide signal transduction pathway through mitogen-activated protein kinase and nuclear factor-kappaB cascades, but not lipopolysaccharide-incorporation steps. Biochim Biophys Acta 1450:130-144

Tully JG, Gaines S, Tigertt WD (1965) Studies on infection and immunity in experimental typhoid fever. VI. Response of chimpanzees to endotoxin and the effect of tolerance on resistance to oral challenge. J Infect Dis 115:445-455

Urbaschek B, Ditter B, Becker KP, Urbaschek R (1984) Protective effects and role of endotoxin in experimental septicemia. Circ Shock 14:209-222

Urbaschek R, Urbaschek B (1985) Induction of nonspecific resistance and stimulation of granulopoiesis by endotoxins and nontoxic bacterial cell wall components and their passive transfer. Ann N Y Acad Sci 459:97-110

van Furth R, van Zwet TL, Buisman AM, and van Dissel JT (1994) Anti-tumor necrosis factor antibodies inhibit the influx of granulocytes and monocytes into an inflammatory exudate and enhance the growth of Listeria monocytogenes in various organs. J. Inf. Dis. 170:234-237

Van Molle W, Denecker G, Rodriguez I, Brouckaert P, Vandenabeele P, Libert C (1999) Activation of caspases in lethal experimental hepatitis and prevention by acute phase proteins. J Immunol 163:5235-5241

Van Molle W, Libert C, Fiers W, Brouckaert P (1997) Alpha 1-acid glycoprotein and alpha 1-antitrypsin inhibit TNF-induced but not anti-Fas-induced apoptosis of hepatocytes in mice. J Immunol 159:3555-3564

Villa P, Shaklee CL, Meazza C, Agnello D, Ghezzi P, Senaldi G (1998) Granulocyte colony-stimulating factor and antibiotics in the prophylaxis of a murine model of polymicrobial peritonitis and sepsis. J Infect Dis 178:471-477

Virca GD, Kim SY, Glaser KB, Ulevitch RJ (1989) Lipopolysaccharide induces hyporesponsiveness to its own action in RAW 264.7 cells. J Biol Chem 264:21951-21956

Vogel SN, Johnson D, Perera PY, Medvedev A, Lariviere L, Qureshi ST, Malo D (1999) Cutting edge: functional characterization of the effect of the C3H/HeJ defect in mice that lack an Lpsn gene: in vivo evidence for a dominant negative mutation. J Immunol 162:5666-5670

Vogel SN, Kaufman EN, Tate MD, Neta R (1988) Recombinant interleukin-1a and recombinant tumor necrosis factor a synergize in vivo to induce early endotoxin tolerance and associated hematopoietic changes. Infect Immun 56:2650-2657

Vogels MT, Eling WM, Otten A, van der Meer JW (1995) Interleukin-1 (IL-1)-induced resistance to bacterial infection: role of the type I IL-1 receptor. Antimicrob Agents Chemother 39:1744-1747

Vogels MT, Hermsen CC, Huys HL, Eling WM, van der Meer JW (1994a) Roles of tumor necrosis factor alpha, granulocyte-macrophage colony- stimulating factor, platelet-activating factor, and arachidonic acid metabolites in interleukin-1-induced resistance to infection in neutropenic mice. Infect Immun 62:2065-2070

Vogels MT, Mensink EJ, Ye K, Boerman OC, Verschueren CM, Dinarello CA, van der Meer JW (1994b) Differential gene expression for IL-1 receptor antagonist, IL-1, and TNF receptors and IL-1 and TNF synthesis may explain IL-1-induced resistance to infection. J Immunol 153:5772-5780

Vogels MT, Sweep CG, Hermus AR, van der Meer JW (1992) Interleukin-1-induced nonspecific resistance to bacterial infection in mice is not mediated by glucocorticosteroids. Antimicrob Agents Chemother 36:2785-2789

Volk H-D, Reinke P, Krausch D, Zuckermann H, Asadullah K, Müller JM, Döcke W-D, Kox WJ (1996) Monocyte deactivation - rationale for a new therapeutic strategy in sepsis. Intensive Care Med 22:474-481

Volk H-De (1991) Alterations in function and phenotype of monocytes from patients with septic disease - predictive value and new therapeutic strategies. Behring Inst Mitt 88:208-215

Waage A, Bakke O (1988) Glucocorticoids suppress the production of tumour necrosis factor by lipopolysaccharide-stimulated human monocytes. Immunology 63:299-302

Wakabayashi G, Cannon JG, Gelfand JA, Clark BD, Aiura K, Burke JF, Wolff SM, Dinarello CA (1994) Altered interleukin-1 and tumor necrosis factor production and secretion during pyrogenic tolerance to LPS in rabbits. Am J Physiol 267:R329-336

Wallach D, Holtmann H, Engelmann H, Nophar Y (1988) Sensitization and desensitization to lethal effects of tumor necrosis factor and IL-1. J Immunol 140:2994-2999

West MA, LeMieur T, Clair L, Bellingham J, Rodriguez JL (1997) Protein kinase C regulates macrophage tumor necrosis factor secretion: direct protein kinase C activation restores tumor necrosis factor production in endotoxin tolerance. Surgery 122:204-211; discussion 211-202

West MA, Li MH, Seatter SC, Bubrick MP (1994) Pre-exposure to hypoxia or septic stimuli differentially regulates endotoxin release of tumor necrosis factor, interleukin-6, interleukin-1, prostaglandin E2, nitric oxide, and superoxide by macrophages. J Trauma 37:82-89; discussion 89-90

Williams Z, Hertogs CF, Pluznik DH (1983) Use of mice tolerant to lipopolysaccharide to demonstrate requirement of cooperation between macrophages and lymphocytes to generate lipopolysaccharide-induced colony-stimulating factor in vivo. Infect Immun 41:1-5

Wilson CS, Seatter SC, Rodriguez JL, Bellingham J, Clair L, West MA (1997) In vivo endotoxin tolerance: impaired LPS-stimulated TNF release of monocytes from patients with sepsis, but not SIRS. J Surg Res 69:101-106

Wise WC, Cook JA, Halushka PV (1983) Arachidonic acid metabolism in endotoxin tolerance. Adv Shock Res 10:131-142

Wittmann M, Larsson VA, Schmidt P, Begemann G, Kapp A, Werfel T (1999) Suppression of interleukin-12 production by human monocytes after preincubation with lipopolysaccharide. Blood 94:1717-1726

Wolk K, Docke WD, von Baehr V, Volk HD, Sabat R (2000) Impaired antigen presentation by human monocytes during endotoxin tolerance. Blood 96:218-223

Yamamoto C, Yoshida S, Taniguchi H, Qin MH, Miyamoto H, Mizuguchi Y (1993) Lipopolysaccharide and granulocyte colony-stimulating factor delay neutrophil apoptosis and ingestion by guinea pig macrophages. Infect Immun 61:1972-1979

Yang RB, Mark MR, Gray A, Huang A, Xie MH, Zhang M, Goddard A, Wood WI, Gurney AL, Godowski PJ (1998) Toll-like receptor-2 mediates lipopolysaccharide-induced cellular signalling. Nature 395:284-288

Yao Z, Foster PA, Gross GJ (1994) Monophosphoryl lipid A protects against endotoxic shock via inhibiting neutrophil infiltration and preventing disseminated intravascular coagulation. Circ Shock 43:107-114

Young HA, Hardy KJ (1995) Role of interferon-gamma in immune cell regulation. J Leukoc Biol 58:373-381

Yoza B, LaRue K, McCall C (1998) Molecular mechanisms responsible for endotoxin tolerance. Prog Clin Biol Res 397:209-215

Yoza BK, Hu JY, Cousart SL, McCall CE (2000) Endotoxin inducible transcription is repressed in endotoxin tolerant cells. Shock 13:236-243

Zhang FX, Kirschning CJ, Mancinelli R, Xu XP, Jin Y, Faure E, Mantovani A, Rothe M, Muzio M, Arditi M (1999) Bacterial lipopolysaccharide activates nuclear factor-kappaB through interleukin-1 signaling mediators in cultured human dermal endothelial cells and mononuclear phagocytes. J Biol Chem 274:7611-7614

Zhang X, Morrison DC (1993) Lipopolysaccharide-induced selective priming effects on tumor necrosis factor alpha and nitric oxide production in mouse peritoneal macrophages. J Exp Med 177:511-516

Ziegler-Heitbrock HW, Blumenstein M, Kafferlein E, Kieper D, Petersmann I, Endres S, Flegel WA, Northoff H, Riethmuller G, Haas JG (1992) In vitro desensitization to lipopolysaccharide suppresses tumour necrosis factor, interleukin-1 and interleukin-6 gene expression in a similar fashion. Immunology 75:264-268

Ziegler-Heitbrock HW, Wedel A, Schraut W, Strobel M, Wendelgass P, Sternsdorf T, Bauerle PA, Haas JG, Riethmuller G (1994) Tolerance to lipopolysaccharide involves mobilization of nuclear factor kappa B with predominance of p50 homodimers. J Biol Chem 269:17001-17004

Ziegler-Heitbrock HWL, Frankenberger M, Wedel A (1995) Tolerance to lipopolysaccharide in human monocytes. Immunobiol 193:217-223

Zingarelli B, Chen H, Caputi AP, Halushka PV, Cook JA (1995a) Reorientation of macrophage mediator production in endotoxin tolerance. Prog Clin Biol Res 392:529-537

Zingarelli B, Halushka PV, Caputi AP, Cook JA (1995b) Increased nitric oxide synthesis during the development of endotoxin tolerance. Shock 3:102-108

Zuckerman SH, Evans GF (1992) Endotoxin tolerance: in vivo regulation of tumor necrosis factor and interleukin-1 synthesis is at the transcriptional level. Cell Immunol 140:513-519

Zuckerman SH, Evans GF, Butler LD (1991) Endotoxin tolerance: independent regulation of interleukin-1 and tumor necrosis factor expression. Infect Immun 59:2774-2780

Zuckerman SH, Evans GF, Snyder YM, Roeder WD (1989) Endotoxin-macrophage interaction: post-translational regulation of tumor necrosis factor expression. J Immunol 143:1223-1227

The Structural Basis
of G-Protein-Coupled Receptor Function
and Dysfunction in Human Diseases

T. Schöneberg[#], A. Schulz, and T. Gudermann*

[#]Institut für Pharmakologie, Universitätsklinikum Benjamin Franklin,
Freie Universität Berlin, Thielallee 69-73, 14195 Berlin, Germany
Tel.: +49-30-8445-1865, Fax: +49-30-8445-1818, E-mail: schoberg@zedat.fu-berlin.de
*Institut für Pharmakologie und Toxikologie, Fachbereich Medizin,
Philipps-Universität Marburg, Karl-von-Frisch-Str. 1, 35033 Marburg, Germany

Contents

Reviews of Physiology, Biochemistry,
and Pharmacology, Vol. 144
© Springer-Verlag Berlin Heidelberg 2002

1
Introduction

Among the different families of transmembrane receptors, G-protein-coupled receptors (GPCR) form the largest superfamily being present in yeast, plants, protozoa and metazoa. Signals as multiform as light, cations, small molecules including ions, amines, amino acids, peptides, lipids, sugars, as well as large proteins enable cells of a multicellular organism to communicate with each other and with their environment. Based on the now entirely known human genome careful estimation suggests that about 3–4% of the human genes code for GPCRs. Upon interaction with extracellular ligands or light activation, GPCRs transduce the signal into the cell by activating a cascade which is initiated by catalyzing GDP-GTP exchange on heterotrimeric G proteins. Besides this main signal transduction pathway recent studies suggest that GPCRs do not only interact with G proteins but are also able to interact with novel receptor-associated proteins mediating and modifying their function.

Because of their central role in controlling almost all physiological functions, accumulating evidence highlights the involvement of GPCRs in many pathophysiological processes. Activating and inactivating mutations in GPCR genes are responsible for an increasing number of human diseases including malignancies. Functional variability resulting from GPCR polymorphisms may contribute to interindividual differences in responses to endogenous and exogenous ligands as well as drugs.

The following review will summarize current knowledge relevant to understanding the molecular basis of GPCR function and focus on the underlying mechanisms of GPCR malfunction responsible for different human diseases. This will provide the basis to discuss strategies aiming at the therapy of diseases caused by receptor dysfunction.

2
Structural Architecture of GPCRs

The diversity of receptor groups within the GPCR superfamily is the result of a long evolutionary process. It has been suggested that serotonin (5-HT) receptors have existed for more than 750 million years (Peroutka and Howell 1994). The tendency toward protein diversification depends upon gene duplications and the continuous accumulation of mutations. The maintenance of vital functions in organisms, however, strictly requires enough structural conservation to ensure the functionality of the corresponding proteins. To

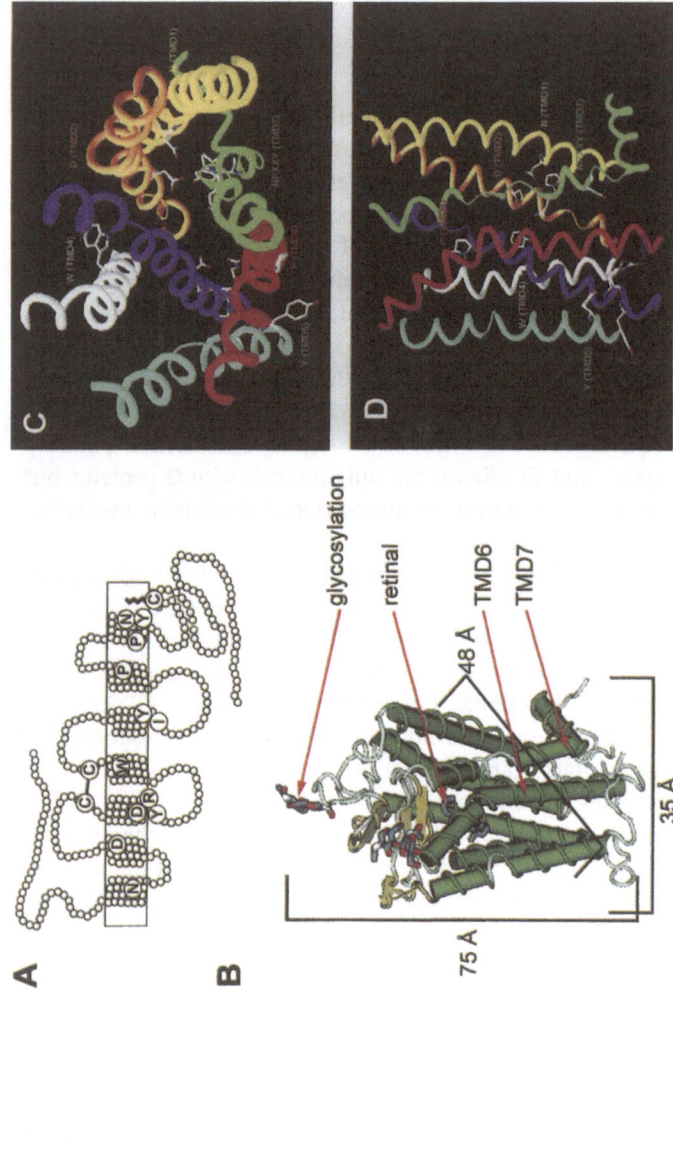

Fig. 1A–D. Structure of family 1 GPCRs. A two-dimensional model of family 1 GPCRs and some of the highly conserved residues are shown in enlarged circles (**A**). The ring-like arranged seven transmembrane helices (TMDs) assemble in a counterclockwise fashion as viewed from the extracellular surface forming a compact receptor structure with a size of 75x35x48 Å (**B**). The models of rhodopsin were generated with the softwares Cn3D (The National Center for Biotechnology Information, version 3.0) and Swiss-PdbViewer (Glaxo Wellcome Experimental Research, version 3.7b2) based on the crystal structure data (Palczewski et al. 2000). The crystal structure of rhodopsin (2.8 Å resolution) shows the orientation of the TMDs relative to each other. Positions of key residues are indicated in the three-dimensional rhodopsin structure viewed from extracellular (**C**) or laterally (**D**)

predict and understand the functional consequences of structural changes within a receptor molecule, detailed information about the native receptor structure in its inactive and active conformation is required. Despite the remarkable structural variety of natural GPCR agonists and a low amino acid sequence homology, hydropathy analysis and biochemical data suggest that all GPCRs share a common molecular architecture consisting of seven transmembrane domains (TMDs). Currently, a high-resolution structure is available only for bovine rhodopsin (Palczewski et al. 2000) because of the difficulties inherent in producing, purifying, and crystallizing other GPCRs. As shown in Fig. 1, the mostly α-helical TMDs are arranged in a closely packed bundle forming the transmembrane receptor core. The seven TMDs of rhodopsin vary in length from 20 to 33 residues. The N terminus of the polypeptide is located in the extracellular space whereas the C terminus shows an intracellular localization. The seven TMDs are connected by six alternating intracellular (i1–i3) and extracellular (e1–e3) loops. An extensive analysis of about 200 GPCR sequences revealed that the total length of GPCRs can vary between 311 and ~1490 amino acid residues. The largest variations in length are found in the N and C termini with sizes up to 879 and 371 amino acid residues, respectively. But also the e2 loops and i3 loops show major size differences of almost 200 residues (Otaki and Firestein 2001).

2.1
GPCR Classification

Molecular cloning studies and genome data analysis have revealed about 1200–1300 members of the GPCR superfamily in the human genome. About 40–60% out of all human GPCRs have orthologs in other species including more distantly related organisms such as *Caenorhabditis elegans* and *Drosophila melanogaster*. To date, about 190 GPCRs have been assigned to an agonist or potential ligands. More than 900 are olfactory GPCRs but the sequence of at least 63% is disrupted in man by what appears to be a random process of pseudogene formation (Glusman et al. 2001). The remaining GPCRs are pseudogenes or so called 'orphan' receptors in man. 'Orphan' GPCRs are cloned GPCRs that bind unknown ligands. More than 200 'orphan' GPCRs, not including the olfactory GPCRs, have been discovered so far. In most cases, the extent of sequence homology is insufficient to assign these 'orphan' receptors to a particular receptor subfamily. Once the sequence of a GPCR is known, understanding the function of the encoded protein becomes a task of paramount importance. Consequently, reverse molecular pharmacological and functional genomic strategies are being employed to identify the activating ligands of the cloned receptors. The reverse molecular pharmacological methodology includes expression of orphan GPCRs in mammalian cells and screening these cells for a functional response to cognate or surrogate agonists present in biological extract preparations or peptide and compound libraries (Debouck and Metcalf 2000). Many new transmitter/receptor systems have been discovered recently, and their physiological functions and potential relevance in human disease are currently being analyzed.

2.1.1
The GPCR Families

The GPCR superfamily comprises at least **three major families** which share little sequence homology among each other. About ninety percent out of all GPCRs are grouped into the rhodopsin-like family (**family 1**). Subclassifying these family 1 receptors is a difficult problem because the mean pairwise amino acid identity is 17%. Family 1 contains a large collection of receptors for autocrine, paracrine, and endocrine factors which include acetylcholine, catecholamine, peptides, glycoprotein hormones, eicosanoids, proteases, nucleotides, and lysosphingolipids. Large extracellular domains (ecto-domain) are rare in GPCRs of family 1. One example comes from the glycoprotein hormone receptors in which the ectodomain is composed of leucine-rich repeats. Recent studies indicated the evolution of an expanding group of homologous leucine-rich repeat-containing GPCRs in family 1 (Hsu et al. 2000).

The GPCR **family 2** (secretin receptor family), the second largest family, recruits about 60 members and is characterized not only by the lack of the structural signature sequences present in the family 1 but also by the presence of a large N-terminal ectodomain. Family 2 comprises receptors for peptides and proteins such as secretin, glucagon, calcitonin, growth hormone-releasing hormone, corticotropin-releasing factor, and pituitary adenylate cyclase-activating peptide. The long and complex disulfide-bonded amino-terminal ectodomain of these receptors plays an important role in agonist binding. Six cysteine residues within the N terminus, two cysteine residues connecting the e1 and e2 loops and about a dozen residues within the TMD core are well conserved among members of this family (Ulrich et al. 1998). Recently cloned GPCRs such as the α-latrotoxin receptor and Ig-Hepta reveal sequences with similarity to family 2 GPCRs within the TMD region but are unusual in so far as they contain large and complex extracellular domains those forming a subfamily within family 2. The first reports of sequences related to family 2 GPCRs followed the isolation of cDNA clones encoding EMR1 (EGF-module-containing mucin-like hormone receptor 1), F4/80, and CD97. The ectodomains are composed of various protein domains such as EGF domains, cadherin repeats, and thrombospondin type-1 repeats (reviewed in Stacey et al. 2000). Although the physiological functions of this subfamily are largely unknown, the acquisition of such extracellular domains leads to the possibility that members of this subfamily possess cell migration and adhesion properties. The *Drosophila* mutant *methuselah* (*mth*) was identified from a screen for single gene mutations that extended average lifespan. The protein affected by this mutation is closely related to GPCRs of the secretin receptor family. The recently resolved 2.3 Å-resolution crystal structure of the *mth* ectodomain shows a three-domain architecture (West et al. 2001).

Family 3 recruits about two dozens GPCRs such as metabotropic glutamate receptors (mGluR), taste receptors, the calcium-sensing receptor, and GABA$_B$ receptors but also potential taste, pheromone and olfactory receptors. Like in family 2, these receptors possess large ectodomains responsible for ligand binding. Family 3 GPCRs are defined as a group of receptors comprising at least three different subfamilies that share $\geq 20\%$ amino acid identity over their seven membrane-spanning regions. Subfamily I includes the metabotropic glutamate receptors, mGluRs 1–8, which are receptors for the excitatory neurotransmitter glutamate and are widely expressed in the central nervous system. Subfamily II contains two types of receptors: the calcium-sensing receptor and a recently discovered, multigene subfamily of putative pheromone receptors. Subfamily III includes a subfamily of receptors, the GABA$_B$ receptors, that bind and are activated by the inhibitory neurotransmitter GABA.

Finally, three additional families encompass pheromone receptors, *Dictyostelium* cAMP receptors, and proteins of the frizzled/smoothened receptor group. These receptors display a heptahelical transmembrane organization but for most of these receptors the terminology 'G-protein-coupled receptor' is controversially discussed because the relevance of G-protein coupling for receptor's signal transduction was not convincingly demonstrated.

GPCRs are not only encoded by eukaryotic genes but also by viral genes. To date, about 20 putative GPCRs have been identified within herpes, pox, and retroviruses (Bai et al. 1999, Tulman et al. 2001). Involvement of GPCRs in the pathophysiologic role of viruses has been impressively demonstrated for the Kaposi's sarcoma-associated herpes virus (KSHV) receptor and the so-called UL78 gene family found in the cytomegalovirus (Arvanitakis et al. 1997, Oliveira and Shenk 2001).

2.1.2
GPCR Diversity Due to Alternative and Tissue-Specific Splicing

Recent molecular characterization of cloned protein genes draws attention to alternative splicing as a source of structural and functional diversity. An astonishing example for gene product multiplication comes from sensory-receptor cells in the inner ear of birds where cell-specific expression of a subset of 576 possible alternatively spliced forms of K^+ channel mRNA occurs (Black, 1998). The amino acid sequences of many GPCRs are encoded by intronless single-copy genes. However, a number of GPCR genes show an exon/intron assembly of their coding regions, as described for rhodopsin, some amine and peptide receptors, and the glycoprotein hormone receptors. In the latter receptor subfamily, only the gene regions encoding the large N terminus are comprised of exons and introns. As a consequence, naturally occurring splice variants have been described for numerous GPCRs. Interestingly, the genomic intron/exon structure and the number of receptor subtypes are not necessarily conserved among species for a given GPCR, e.g. the angiotensin receptor subtypes in human and mouse. Extensive studies on the rhodopsin gene have shown that introns in the coding region can appear and disappear during evolution. For example, human rhodopsin is encoded by four exons, but in some fish species the coding region for rhodopsin is intronless (Venkatesh et al. 1999). The existence of introns in GPCR genes provides the potential for additional diversity by virtue of alternative splicing events which may generate distinct receptor isoforms. For example, pharmacological and molecular biological studies have resulted in the cloning of cDNAs encoding four EP prostanoid receptors. The cloning of these receptors has revealed further heterogeneity due to alternative mRNA splicing. Specifically, eight human EP_3 receptor isoforms have been identified which differ only in their C termini (Pierce and Regan, 1998). It should be noted that a tissue-specific occurrence of distinct splice variants has been described, e.g. for the PACAP receptor (Chatterjee et al. 1996) and the corticotropin releasing factor (CRF) receptor (Ardati et al. 1999).

In principle, **two types of GPCR splice variants** can be distinguished. First, usage of an alternative splice site can generate a functional receptor as demonstrated for a large number of GPCRs such as the EP_3 receptor (Namba et al. 1993), the thromboxane receptor (Vezza et al. 1999), the metabotropic glutamate receptor 1 (Prezeau et al. 1996), and the D_2 dopamine receptor (Guiramand et al. 1995, Seeman et al. 2000). Most variations are found in the i3 loop and the receptor C terminus which are considered to be important for G-protein coupling and interactions with other proteins (see 3.2). As shown for the EP_3 and $5-HT_4$ receptor isoforms, alternative splice products can vary in their basal activity when expressed *in vitro*. The extent of constitutive activity was found to be reversally correlated with the length of the C-terminal portion of the splice variants (Jin et al. 1997, Claeysen et al. 1999). In most cases, the divergence between receptor isoforms is limited to the C-terminal tail, a region involved in internalization and down-regulation. In contrast, some GPCR splice variants, when expressed *in vitro*, displayed similar pharmacological profiles and signaling specificity (Ito et al. 1994, Park et al. 2000). Second, an improper splicing event can produce a new protein that may display dominant negative effects on the wild-type receptor. It has been demonstrated that the expression of a truncated form of the gonadotropin-releasing hormone (GnRH) receptor can decrease the signaling efficacy of the full length receptor by reducing its cell surface expression levels (Grosse et al. 1997). This dominant negative effect was highly specific for the GnRH receptor and was probably due to heterocomplex formation between the two proteins. One may speculate that co-expression of truncated receptor isoforms may modulate the gonadotropes' responsiveness to GnRH and thus contribute to the fine tuning of gonadotropin release *in vivo*. Similarly, the ability of an EP_1 receptor isoform to inhibit signaling by EP_1 as well as EP_4 receptors can be explained by complex formation between these different receptors (Okuda-Ashitaka et al. 1996).

2.1.3
GPCR Diversity Due to RNA Editing

RNA editing is a co- or post-transcriptional process in which selected nucleotide sequences in the mRNA are altered when compared to the genomic sequence. Double-stranded RNA-specific adenosine deaminases convert adenosine residues to inosine in messenger RNA precursors (pre-mRNA). Their main physiological substrates are pre-mRNAs. Extensive analysis of cDNAs from 5-HT_{2C} receptor reveals posttranscriptional modifications indicative of adenosine-to-inosine RNA editing (Burns et al. 1997). RNA transcripts encoding the 5-HT_{2C} receptor undergo adenosine-to-inosine RNA editing events at up to five specific sites. Interestingly, reduced G-protein-coupling efficiency for the edited isoforms is primarily due to silencing of the constitutive activity of the non-edited 5-HT_{2C} receptor (Niswender et al. 1999). No further example of modified GPCR functions by mRNA editing has been reported yet.

2.2
Arrangement and Structure of the Transmembrane Domains

The understanding of GPCR function and dysfunction requires detailed information on the receptor core structure. Like other polytopic membrane proteins, GPCRs are partially buried in the non-polar environment of the lipid bilayer by forming a compact bundle of transmembrane helices. The correct orientation and integration of the polypeptide chain is guided by a complex translocation apparatus residing in the endoplasmic reticulum (ER). Two different folding stages can be distinguished following an initial translocation of the receptor N terminus into the ER lumen. In stage I, hydrophobic α-helices are established across the lipid bilayer, and protein folding is predominantly driven by the hydrophobic effect. The TMDs adopt a secondary structure in order to minimize the polar surface area exposed to the lipid environment with the result that hydrophobic amino acids face the lipid bilayer and that the more hydrophilic amino acid residues are orientated towards the core crevice of the TMD bundle. In stage II, a functional tertiary structure is formed by establishing specific helix-helix interactions, leading to the tightly packed, ring-like structure of the TMD bundle.

In the early stage of GPCR structure/function analysis, investigators used the structure of bacteriorhodopsin, a prokaryotic ion pump with structural similarities to the GPCR superfamily, as a scaffold for topographical models of the transmembrane core of GPCRs (Baldwin, 1994). The identification of specific interhelical contact sites was required to provide information about the relative orientation of the different helices towards each other. To determine the structural determinants which actually contribute to specific helix-helix interactions, chimeric receptors were generated. Studies with chimeric muscarinic receptors provided the first experimental evidence as to how TMD1 and TMD7 are oriented relative to each other and also strongly suggested that the TMD helices in muscarinic receptors are arranged in a counterclockwise fashion as viewed from the extracellular membrane surface (Liu et al. 1995). Functional analysis of artificial metal ion-binding sites (Elling et al. 1995) and disulfide bonds (Farrens et al. 1996) as well as spectroscopic approaches (Beck et al. 1998) allowed the identification of distinct amino acid residues that are involved in helix-helix contacts and the relative orientation of single TMDs to each other. Nuclear magnetic resonance (NMR) and circular dichroism (CD) studies with peptides derived from the cytoplasmic domains of GPCRs predicted a cytosolic α-helical extension of all TMDs (Jung et al. 1995, Yeagle et al. 1997, Schulz et al. 2000a). This assumption is experimentally supported by site-directed spin-labeling studies (Farahbakhsh et al. 1995, Altenbach et al. 1996) and mutagenesis studies (Biebermann et al. 1998). Finally, the proposed helical arrangement was confirmed by low resolution structures of the transmembrane core of rhodopsin (Unger et al. 1997) and the rhodopsin crystal structure (Palczewski et al. 2000) highlighting the feasibility of mutagenesis approaches and structural analyses of single TMD fragments. Interestingly, the overall α-helical character of TMDs is often disrupted by non-α-helical components, such as intrahelical kinks (often due to residues other than proline), 3_{10}-helices and π-helices (Riek et al. 2001).

2.3
Conserved Structural Features and their Functional Relevance

Most of the current knowledge about structure/function relationships of GPCRs is based on studies with rhodopsin and other members of the family 1 of GPCRs. Thus, this section will mainly focus on data obtained with rhodopsin-like receptors. Only a few critical amino acid residues have been preserved during evolution of the rhodopsin-like GPCR family (see Fig. 1). Despite an evolutionary conservation mutational alteration of some conserved amino acid residues does not always have the same functional consequence.

2.3.1
Disulfide Bonds in GPCRs

The majority of family 1 and also family 2 GPCRs contains a conserved pair of extracellular cysteine residues linking the first and second extracellular loops via a disulfide bond (Hausdorff et al. 1990). Numerous functional analyses of mutant GPCRs in which the cysteine residues were replaced by other amino acids have shown that this disulfide bond may be critical for receptor signaling (Kosugi et al. 1992, Savarese et al. 1992, Cook and Eidne 1997). In some receptors, however, this disulfide bond is required to maintain more distinct functions. Systematic mutagenesis studies of the conserved cysteine residues in several GPCRs showed that disruption of the disulfide bond does not influence the receptor's ability to activate G proteins but interferes with high affinity ligand binding and receptor trafficking (Le Gouill et al. 1997, Perlman et al. 1995, Schulz et al. 2000b, Zeng et al. 1999). Despite the disruption of the disulfide bond, the receptor core structure appears to remain intact, allowing receptor function. Consistent with this notion, some GPCRs, e.g. receptors for sphingosine 1-phosphate and lysophosphatidic acid, lack the conserved extracellular Cys residues. Interestingly, many GPCRs including receptors for biogenic amines, peptides and many 'orphan' GPCRs contain a second conserved pair of extracellular cysteine residues linking the N terminus and third extracellular loop. Mutational disruption of this disulfide bond results in a loss of high affinity binding of P2Y$_1$ receptor ligands, suggesting a pivotal role of an N terminus/e3 loop-connecting disulfide bridge for proper receptor assembly (Hoffmann et al. 1999). In chemokine receptors this disulfide bond is required for ligand binding and agonist-induced receptor function but not for constitutive activity of KSHV-GPCR (Ho et al. 1999) and CCR$_5$-mediated HIV entry (Blanpain et al. 1999). In the crystal structure of rhodopsin, the N-terminal segment is located just below the e3 loop. Specific non-covalent contacts maintain the proper orientation between the rhodopsin N terminus and the extracellular loops so that an additional disulfide bridge like in other GPCRs is probably not required.

2.3.2
Conserved Amino Acid Residues and Motifs in Family 1 GPCRs

Within their transmembrane core most family 1 GPCRs possess a number of highly conserved residues, such as an Asp residue in TMD2, a DRY motif at the TMD3/i2-transition, a Trp residue in TMD4, a Tyr residue in TMD5, a Pro residue in TMD6 and an N/DP(X)$_n$Y motif in TMD7 (see Fig. 1). For example, the **DRY motif** located at the boundary of TMD3 and the i2 loop is a highly conserved triplet of amino acid residues known to play an essential role in GPCR function. The crystal structure of rhodopsin proposes that the residues of the DRY motif participate in several hydrogen bonds with surrounding residues of TMD6 (Palczewski et al. 2000). It was shown for rhodopsin that the corresponding Glu residue (ERY motif) is involved in proton uptake resulting in the formation of the activated metarhodopsin II intermediate (Helmreich and Hofmann, 1996). Similarly, the Asp residue in the DRY motif has been proposed to act as a proton acceptor during receptor activation, as shown for the α_{1B} adrenergic receptor (Scheer et al. 1997). Mutation of this Asp residue results in constitutive activity of many receptors (Alewijnse et al. 2000, Rasmussen et al. 1999). Kobilka and co-workers found that protonation increases basal activity by destabilizing the inactive state of the ß$_2$ adrenergic receptor but the pH sensitivity of receptor activation was not abrogated by mutation of Asp130, which is homologous to the conserved acidic amino acid residue in the DRY motif of rhodopsin and the α_{1B} adrenergic receptor (Ghanouni et al. 2000). In the crystal structure the Glu residue forms a salt bridge with the Arg residue of the ERY motif. However, several family 1 GPCRs are known in which the acidic residue (Asp, Glu) within this motif is naturally substituted by His, Asn, Gln, Gly, Val, Thr,

Cys or Ser residues, questioning the general importance of a salt bridge and a proton uptake at this amino acid position for GPCR activation. The fully conserved Arg residue in the DRY motive is considered to be a key residue in signal transduction of GPCRs. Replacement of the conserved Arg residue by different amino acids virtually abolished G-protein coupling of many GPCRs (Franke et al. 1992, Jones et al. 1995, Scheer et al. 1996, Zhu et al. 1994). Therefore, the conserved Arg residue has been implicated as a central trigger of GDP release from the G-protein α subunit (Acharya and Karnik 1996). Recent studies with mutants of the N-formyl peptide receptor, luteinizing hormone receptor (LHR), and V2 vasopressin receptor (AVPR2) showed that G-protein coupling is only decreased but not abolished after replacement of the Arg residue in the DRY motif (Arora et al. 1997, Schöneberg et al. 1998, Seibold et al. 1998, Schulz et al. 1999) probably due to reduced receptor cell surface expression levels in response to constitutive arrestin-mediated desensitization (Barak et al. 2001). The conserved Arg residue in the human β_2 adrenergic receptor, the TRH receptor, and the CB_2 cannabinoid receptor could also be substituted by other amino acids without loss of G-protein coupling, indicating that the presence of the conserved Arg residue is not an absolute requirement for G-protein activation (Perlman et al. 1997, Seibold et al. 1998, Rhee et al. 2000). Taking advantage of the structurally stabilizing effect of ligands, the impaired cell surface expression and, therefore, the signal transduction of the H_2 histamine receptor mutant R116A/N was partially restored by preincubation with either an agonist or inverse agonist (Alewijnse et al. 2000). Therefore, the GPCR dysfunction caused by replacement of the Arg within the DRY motif is caused by improper receptor folding and trafficking rather than by a direct effect on receptor/G-protein coupling efficacy in many GPCRs. These results can be reconciled by assuming that the DRY motif has different functions in different GPCR classes. It has been speculated that the Arg residue directly interacts with a conserved Glu/Asp residue at the very N-terminal end of TMD6 by forming a salt bridge in the inactive conformation of the β_2 adrenergic receptor (Ballesteros et al. 2001) but refinement of the rhodopsin crystal structure (Teller et al. 2001) as well as functional studies with glycoprotein hormone receptors do not support such interaction (Schulz et al. 2000a).

The **N/DP(X)$_n$Y motif** within the TMD7 near the cytoplasmic face of the plasma membrane is highly conserved (see Fig. 1). This sequence has been postulated to play important roles in receptor activation and regulation (Wang et al. 1996). It was shown that light-activation of rhodopsin rendered an epitope including residues of the N/DP(X)$_n$Y motif accessible for an epitope-specific antibody, suggesting conformational changes of this sequence motif (Abdulaev and Ridge 1998). The Asn residue in this sequence motif is thought to play a crucial role in receptor activation. In the cholecystokinin type B receptor, mutation of the Asn in the N/DP(X)$_n$Y motif to Ala had no effect on cell surface expression and high affinity ligand binding but completely abolished G$_q$-mediated signaling (Gales et al. 2000). Mutational alteration of the conserved Pro residue in the N/DP(X)$_n$Y motif resulted in a complete loss of receptor function as demonstrated *in vivo* for the AVPR2 (Tajima et al. 1996) underlining the functional importance of this motif. It is noteworthy that in a few GPCRs of family 1 the Asn/Asp residue within the N/DP(X)$_n$Y is naturally replaced by Ser, Thr, Lys, or His.

2.3.3
Posttranslational Modifications of GPCRs

The polypeptide chain of most GPCRs is posttranslationally modified including glycosylation, palmitoylation and phosphorylation. Potential N-glycosylation sites (NXS/T) and O-glycosylation sites (Sadeghi and Birnbaumer 1999, Nakagawa et al. 2001) are usually located within the extracellular N-terminal region but are also found in the extracellular loops. The number and exact positions of glycosylation sites are usually not conserved among orthologs of different species. Some GPCRs, e.g. the A$_2$ adenosine and the human α_{2B} adrenergic receptors, completely lack consensus sites for glycosylation in their N termini but are fully active in the absence of this posttranslational modification. The functional relevance of post-translational modifications in GPCRs has been extensively studied in *in vitro* systems. It is well accepted that mutational disruption of potential N-glycosylation sites of most GPCRs has little effect on receptor function *in vitro* (Rands et al. 1990, Innamorati et al. 1996, Zeng et al. 1999). However, nonglycosylated receptors for parathyroid hormone and glycoprotein hormones in which glycosylation sites were mutated, are deficient in function (Zhang et al. 1995, Zhou et al. 2000). In the human calcium-sensing receptor, eight out of 11 potential N-linked glycosylation sites are actually utilized. Glycosylation of at least three sites is critical for cell surface expression of the receptor, but glycosylation does not appear to be critical for signal transduction (Ray et al. 1998).

Consensus acceptor phosphorylation sites for protein kinases A and C and potential receptor-specific kinase phosphorylation sites (multiple serine and threonine residues) are present in the i3 loop and the C-terminal domain. Several studies indicated that the selectivity of receptor/G-protein coupling is regulated by receptor phosphorylation (3.1.2).

Most GPCRs contain one or more conserved cysteine residues within their C-terminal tail (see Fig. 1) which are modified by covalent attachment of palmitoyl or isoprenyl residues (Bouvier et al. 1995, Hayes et al. 1999). As known from the rhodopsin structure the palmitoyl moiety is anchored in the lipid bilayer forming a fourth intracellular (i4) loop. Depending on the specific GPCR examined, different effects on receptor phosphorylation, internalization, trafficking and G-protein-coupling profile have been described (reviewed in Bouvier et al. 1995, Wess 1998). However, several family 1 GPCRs do not have Cys residues in their C-terminal tail for posttranslational modification probably lacking the i4 loop.

2.4
GPCR Assembly and Oligomerization

2.4.1
GPCR Assembly from Independent Folding Units

Following pioneering studies with bacteriorhodopsin (Huang et al. 1981, Kahn and Engelman 1992), the successful reconstitution of adrenergic receptors (Kobilka et al. 1988) from two fragments demonstrated that the integrity of the GPCR polypeptide chain is not required for proper receptor function. Based on these findings it has been speculated that GPCRs are composed of two or more independent folding domains. To test this hypothesis rhodopsin and the m3 muscarinic receptor were split in all three intracellular and extracellular loops. It was shown that except for a construct containing only TMD1, a significant portion of all N- and C-terminal receptor fragments studied was found to be inserted into the plasma membrane in the correct orientation even when expressed alone (Ridge et al. 1995, Schöneberg et al. 1995). Co-expression of some complementary receptor polypeptide pairs, generated by splitting GPCRs in their intra- and extracellular loops, resulted in receptors which were able to bind ligands and to mediate agonist-induced signal transduction (reviewed in Gudermann et al. 1997). It is noteworthy that all attempts to assemble functional receptor proteins from solubilized receptor fragments *in vitro* were unsuccessful (Schöneberg et al. 1997). This indicates that molecular chaperones such as RanBP2, nina A and calnexin that are likely to assist folding of the wild-type receptor protein may also play a role in facilitating complex formation. It has been shown that chaperone-dependent mechanisms are essential for proper folding of rhodopsin (Baker et al. 1994, Ferreira et al. 1996), AVPR2 (Morello et al. 2001) and gonadotropin receptors (Rozell et al. 1998).

2.4.2
GPCR Multimerization

The ability of functional complementation from receptor fragments is consistent with reports showing or suggesting that GPCRs can form dimers and oligomers. For several non-GPCR receptor families, such as receptor tyrosine kinases and kinase-associated cytokine receptors, agonist-induced receptor dimerization is required for initiating a signal transduction cascade. First evidence for GPCR dimerization came from crosslinking and photoaffinity labelling experiments with GnRH, LH, and muscarinic receptors (Conn et al. 1982, Avissar et al. 1983, Podesta et al. 1983). Numerous studies describing similar findings followed, but most reports of GPCR di- and oligomerization were based on co-immunoprecipitation studies. It has been argued that biochemical evidence from co-immunoprecipitation and Western blot experiments supporting the existence of GPCR oligomers is questionable, since solubilization of integral transmembrane proteins can cause artificial aggregation. However, as shown for epitope-tagged $ß_2$ adrenergic, muscarinic and vasopressin receptors, the association is highly specific for a given receptor subtype giving rise only to homodimers (Hebert et al. 1996, Zeng and Wess 1999, Schulz et al. 2000b). In addition to investigations in transient expression systems, *in vivo* studies with D_2 and D_3 dopamine receptors (Nimchinsky et al. 1997, Zawarynski et al. 1998), somatostatin receptor type 5 ($SSTR_5$)/D_2 dopamine receptors (Rocheville et al. 2000a), and rhodopsin (Colley et al. 1995) suggest the coexistence of receptor monomers and oligomeric complexes under physiological circumstances.

One question that arises from these studies is as to whether GPCR dimers are pre-formed or are induced in the presence of the appropriate ligand. Most co-immunoprecipitation data suggest the existence of oligomeric receptor complexes under basal conditions. Examining the biological relevance of GPCR homodimerization *in vivo*, Bouvier and co-workers used a bioluminescence resonance energy transfer (BRET) technique to study receptor-receptor interactions. It was shown that $ß_2$ adrenergic receptors and thyrotropin-releasing hormone receptors (TRHR) form constitutive homodimers that are expressed at the cell surface where they interact with agonists (Angers et al. 2000, Kroeger et al. 2001). Constitutive receptor association appears to be a general phenomenon since the yeast α-mating factor receptor forms dimers under basal conditions, as shown by a fluorescence resonance energy transfer (FRET) approach (Overton and Blumer 2000). On the other hand, there is also experimental support for an agonist driven oligomerization mechanism. Thus, the B_2 bradykinin receptor and the

CXCR$_4$ receptor undergo receptor dimerization after ligand binding (AbdAlla et al. 1999, Vila-Coro et al. 1999).

There is growing evidence that GPCR not only exist in homodimeric structures but also in complexes formed by different GPCRs. Expression of the recombinant GABA$_{B1}$ receptor in COS cells resulted in a significantly lower agonist affinity when compared with native receptors. Interestingly, co-expression of the GABA$_{B1}$ receptor and the GABA$_{B2}$ receptor, a recently cloned novel GABA$_B$ receptor subtype, in *Xenopus* oocytes and HEK-293 cells led to efficient coupling to G-protein-regulated inward rectifier K$^+$ channels (GIRKs) with an agonist potency in the same range as for GABA$_B$ receptors in neurons (Jones et al. 1998, White et al. 1998, Kaupmann et al. 1998). Encouraged by these studies, an ever growing number of heterodimeric complexes has been identified. For example, there is biochemical and pharmacological evidence that the κ and δ opioid receptors as well as μ and δ opioid receptors associate with each other. The complexes exhibit ligand binding and functional properties that are distinct from those of either receptor (Jordan and Devi 1999, George et al. 2000). Heterodimer formation was also observed for other receptor subtypes such as 5-HT$_{1B}$/5-HT$_{1D}$ receptors and SSTR$_1$/SSTR$_5$ (Xie et al. 1999, Rocheville et al. 2000b). Hetero-oligomerization between the D$_2$ dopamine receptor and SSTR$_5$ was demonstrated, resulting in a novel receptor with enhanced functional activity (Rocheville et al. 2000a). In a very recent study, Zuker and colleagues showed that formation of a functional sweet taste receptor is only achieved when two taste GPCRs, T1R2 and T1R3, are coexpressed (Nelson et al. 2001). The ability of GPCRs to heterodimerize provides a new mechanism by which a cell can fine-tune its responsiveness to an agonist via co-expression of distinct GPCR subtypes.

2.4.3
The Structure of GPCR Multimers

The molecular mechanisms and structural requirements which are responsible for GPCR oligomerization are only poorly understood. There is evidence for at least **three molecular mechanisms** by which GPCRs can oligomerize – covalent linkage via disulfide bonds, domain interaction such as coiled-coil structure and specific interaction of the TMD regions (Fig. 2). In the case of the mGluR$_5$ (Romano et al. 1996) and the calcium-sensing receptor (Bai et al. 1998, Ward et al. 1998), which are members of family 3, disul-

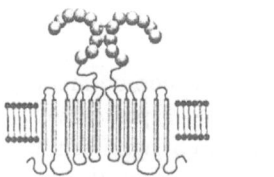 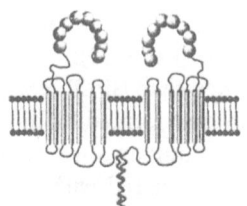 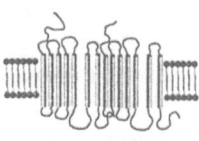

disulfide bond-linked dimer
mGlu receptors
Ca²⁺-sensing receptor

coiled-coil-linked dimer
GABA_B receptors

lateral contact dimer
many family 1 receptors
(homo- and heterodimers)

domain-swapped dimer   lateral contact dimer

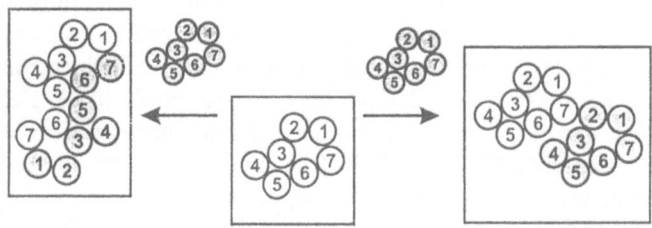

Fig. 2. Hypothetical structures of GPCR dimers. GPCR-dimer formation is can be mediated via covalent (disulfide bond) and non-covalent (coiled-coil structure, lateral contact) interactions (upper panel). The molecular structure of most family 1 homo- and heterodimers is currently unknown. Transmembrane domains (TMDs, numbered from 1 to 7) of GPCRs form a ring-like structure in a counter-clock wise fashion as viewed from extracellular (lower panel). GPCRs are composed of at least two independent folding domains (TMDs1-5 and TMDs6-7) which are connected by the i3 loop. Accumulating evidence suggests that wild-type GPCRs can exist in dimeric complexes, and two structural models of dimer formation have been suggested (Gouldson et al. 1998). The contact interface of so-called 'swapped dimers' is recruited from the rearrangement of two independent folding domains of the individual receptor monomers. The ring-like TMD arrangement is still retained by the complementary exchange of the two folding domains. In lateral contact dimers, a site-to-site interaction of the individual receptor molecules is assumed

fide bonds between the extracellular portions are of critical importance for receptor dimerization. Recently, crystal structures of the extracellular ligand-binding region of mGluR1 – in a complex with glutamate and in two non-liganded forms – have been resolved showing disulfide-linked homodimers (Kunishima et al. 2000). In contrast, it was demonstrated that mutant calcium-sensing receptors without extracellular cysteines form dimers on the cell surface to a similar extent as observed for wild-type receptors (Zhang et al. 2000). Interestingly, the GABA$_{B2}$ receptor was initially discovered by a yeast two hybrid approach using the C terminus of the GABA$_{B1}$ receptor for screening a human brain cDNA library. Heterodimer formation was assumed to be mediated via a coiled-coil structure of the C termini of the two receptors (White et al. 1998). It was found later that a C-terminal retention motif RXR(R) is masked by GABA$_B$ receptor dimerization allowing the plasma membrane expression of the assembled complexes (Margeta-Mitrovic et al. 2000). However, association of both GABA$_B$ receptors was demonstrated even in the absence of their cytoplasmic C termini (Pagano et al. 2001).

Most studies agree that homo- and heterodimers found for rhodopsin-like GPCRs represent non-covalent complexes. Thus, two structural models of dimer formation have been proposed for family 1 GPCRs (Gouldson et al. 1998). In one dimeric structure, referred to as 'contact dimer', two tightly packed bundles of seven TMDs are positioned next to each other. The contact interface between the two monomeric receptors is assumed to be located between the lipid-orientated transmembrane receptor portions (see Fig. 2). The so-called 'domain-swapped dimer' has been proposed to explain the reconstitution phenomenon observed with truncated and chimeric GPCRs (Maggio et al. 1993, Schulz et al. 2000b). In this dimer structure, the two receptor molecules fold around a hydrophilic interface by exchanging their N-terminal (TMDs1–5) and C-terminal (TMDs6–7) folding domains (see Fig. 2). Attempting both hypothetical dimer structures data with the AVPR2 and D$_2$ dopamine receptor strongly support an oligomeric structure in which family 1 GPCRs form contact oligomers by lateral interaction rather than by a domain-swapping mechanism (Schulz et al. 2000b, Lee et al. 2000). High resolution X-ray structure determinations of three heptahelical membrane proteins, the bacteriorhodopsin, the halorhodopsin and the sensory rhodopsin II, clearly show that both proteins assemble to multimers (Luecke et al. 1999, Kolbe et al. 2000, Royant et al. 2001). The proton pump bacteriorhodopsin shares structural similarities with the GPCR family including the assembly from multiple independent folding units. In the trimeric structure found in bacteriorhodopsin and halorhodopsin crystals, TMDs2–4 of the three molecules face each other forming an inner circle of TMDs. In the dimeric sensory rhodopsin II TMD1 and TMD7 contact each other. Structural data did not provide any support for a domain-swapping mechanism of oligomerization. Similarly, other polytopic membrane-spanning proteins which homo-oligomerize in order to build a functional complex, such as aquaporins, assemble via lateral interaction (Walz et al. 1997).

Taken together, recent data strongly support oligomeric GPCR structures. Functional studies with mutant GPCRs provided strong evidence that oligomerization occurs by lateral interaction rather than by a domain-swapped mechanism (Lee et al. 2000, Schulz et al. 2000b). There is growing evidence that GPCR dimerization has consequences for physiologic receptor functions such as formation of receptor 'subtypes' with new ligand binding or signaling abilities (Jordan and Devi 1999, AbdAlla et al. 2000).

3
Diversity of Physiological GPCR Function

In the post-genome era long established views of receptor pharmacology are changing since modifications of GPCR structure and function can contribute to the pharmacological diversity found for products of a single GPCR gene. The changes in GPCR structure and function can occur at different levels. For example, gene duplication events that have led to multiple receptor subtypes are the cause for a considerable functional diversity found in one transmitter system. Moreover, tissue-specific splicing, RNA editing, and variations in posttranslational modifications can multiply the products derived from one GPCR gene. In addition, tissue- or cell-specific expression of effectors can modify the ligand preference and signal transduction capabilities of GPCRs, so that specificity and function of a given GPCR can vary when expressed in a different cellular environment.

3.1
Current Models of Receptor Activation

It is truly one of the most exciting issues within the GPCR field to unravel structural changes in the receptor protein while transducing extracellular signals. The classical 'ternary complex model' postulates that receptor activation leads to the agonist-promoted formation of an active "ternary" complex of agonist, receptor, and G protein (De Lean et al. 1980). This model had to be extended in order to account for the fact that many receptors can activate G proteins in the absence of agonist (Lefkowitz et al. 1993). Based on these seminal observations, receptors are assumed to exist in an equilibrium between the inactive state R and the active state R* (Fig. 3). The model predicts that, even in the absence of agonist, a certain fraction of receptors will spontaneously adopt an active conformation, permitting agonist-independent G-protein activation. In keeping with this current concept, some wild-type receptors like the D_{1B} dopamine receptor, the H_3 histamine receptor, the melanocortin type 4 (MC4) receptor and the thyrotropin recep-

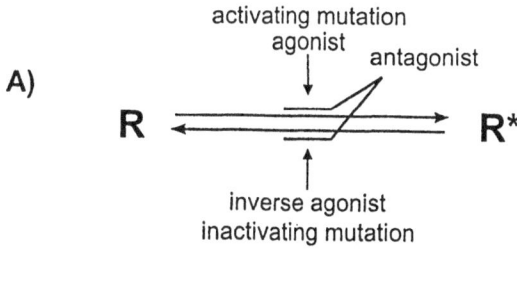

A)

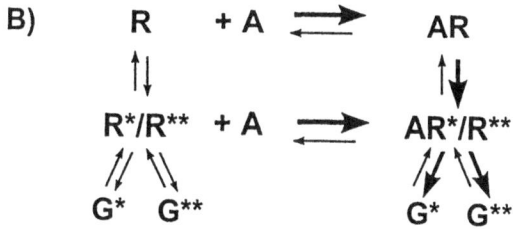

B)

Fig. 3. Ligand- and mutation-induced effects on the functional equilibrium of GPCRs (A) and the functional conformations in GPCRs with multiple coupling abilities (B). (A) According to allosteric ternary complex models of receptor activation (Lefkowitz et al. 1993), GPCRs exist in an equilibrium between the inactive (R) and active (R*) conformation responsible for spontaneous activity. R* has high affinity for G proteins. Agonist (A) binding to the free receptor (uncoupled from G proteins) leads to stabilization of a partly activated form of the receptor (R*) that is able to couple to the G protein. Inverse agonists decrease the spontaneous activity of receptors by shifting the equilibrium towards R. Like agonists or inverse agonists, activating or inactivating mutations are capable of shifting the equilibrium toward R* and R, respectively. Neutral antagonists bind to the receptor but have no influence on the R/R* equilibrium and compete for both agonists and inverse agonists. (B) Agonists also stabilize the activated, coupled form of the receptor (AR* + G*). Most GPCRs couple to more than one G protein (G* and G**). There is evidence that distinct conformational states (R* and R**) favor the interaction with one or the other G-protein family

tor (TSHR) display significantly elevated basal activity even in the unliganded state (Tiberi and Caron 1994, Cetani et al. 1996, Biebermann et al. 1998, Nijenhuis et al. 2001). To describe mutant receptors characterized by a shift of the isomerization equilibrium towards the active conformation, the term 'constitutive activity' has been coined. By definition, such receptors can activate G proteins in the absence of agonist.

3.1.1
Structural Aspects of GPCR Activation

Conformational changes accompanying receptor activation have been initially studied in the photoreceptor rhodopsin. Here, the ligand, 11-*cis*-retinal, is already positioned in the transmembrane bundle and isomerizes upon illumination to function as agonist. Using a variety of biophysical methods, conformational differences between the inactive rhodopsin and the activated metarhodopsin II were characterized (reviewed in Helmreich and Hofmann 1996, Okada et al. 2001). Site-directed spin-labeling studies and studies with artificially created zinc binding sites indicate that movements of TMD3 and TMD6 relative to one another are required for G-protein activation (Farrens et al. 1996, Sheikh et al. 1996). Available evidence indicates that light-induced helix movements in rhodopsin occur only after termination of the early photochemical events. In the subsequent so-called 'Meta'- states, the photoreceptor remains in a G-protein dependent equilibrium between active and inactive conformations and the photolyzed chromophore all-*trans*-retinal or its analogs act as full or partial agonists (Okada et al. 2001). This has provided the evidence that the metarhodopsin states fit into the general ternary complex scheme for GPCRs. Additional evidence for a relative movement between TMD3 and TMD6 in other GPCRs was provided by direct fluorescence labeling of the β_2 adrenergic receptor (Gether et al. 1995, Gether et al. 1997, Ghanouni et al. 2001) or by monitoring the accessibility of Cys residues to a hydrophilic sulfhydryl-specific reagent during receptor activation (Javitch et al. 1997).

Site-directed mutagenesis approaches have been used successfully to examine several aspects of receptor structure-function relationships including ligand binding, G-protein coupling, receptor folding, and mechanisms of activation. *In vitro* mutagenesis studies with several GPCRs provided compelling evidence for the existence of intramolecular constraining determinants which stabilize the inactive receptor conformation. Mutational alteration of such intramolecular contact sites can lead to constitutive receptor activation. Most of the mutations found to be responsible for constitutive receptor activity are located in the C-terminal portion of the i3 loop and within different TMDs. Interhelical salt bridges, as specific structural determinants stabilizing the inactive state, have been identified in rhodopsin (Robinson et al. 1992) and the α_{1B} adrenergic receptor (Porter et al. 1996). The identification of such specific contact sites provides valuable information about the orientation of the different helices relative towards each other.

The largest number of different activating mutations has been identified in the TSHR and LHR, providing important insights into the structural requirements of receptor quiescence. In the LHR, a tightly packed hydrophobic cluster and a specific H-bonding network formed between the cytoplasmic portions of TMD5 and TMD6 and the central regions of TMD6 and TMD7 is thought to maintain the inactive receptor conformation. Mutagenesis data suggested that LHR activation is associated with the disruption of key interhelical side-chain interactions (Lin et al. 1997). Recent data with constitutive active LHR and TSHR mutants implicate that in addition to interhelical interactions the inactive conformation of GPCRs is also stabilized by specific intrahelical structures (Schulz et al. 2000a).

3.1.2
Determinants of G-Protein/Receptor Interaction and Coupling Specificity

Since more than one thousand GPCRs interact with a limited repertoire of G proteins, the issue of coupling specificity needs to be addressed. Based on primary amino acid similarity of the α subunits, G proteins are grouped into four major families, G_s, $G_{i/o}$, $G_{q/11}$, and $G_{12/13}$. Nearly two decades after the cloning of the first GPCRs, there are still many open questions relating to the mechanisms of GPCR/G-protein interaction and the molecular elements determining G-protein coupling specificity. Over the past few years, an increasing number of GPCRs with a broad G-protein coupling profile has been identified (Gudermann et al. 1997).

Structural elements determining signaling specificity are located in both the G protein and the receptor. Numerous *in vitro* mutagenesis studies have been performed with G-protein α subunits to understand how coupling selectivity is achieved. The C terminus as well as the N terminus of the α subunit make important contributions to appropriate receptor/G-protein recognition (Wess, 1998).

The exact nature of G-protein interaction sites within the receptor is currently unknown and may vary between the different GPCRs and G proteins. It is assumed that not only the intracellular loops but also the cytoplasmic sides of the TMDs participate in GPCR/G-protein coupling. Indeed, peptides derived from the i3 loop/TMD6 junction can activate G proteins (Abell and Segaloff 1997, Varrault et al. 1994). Likewise, site-directed mutagenesis studies with GPCRs examined the structural elements that participate in G-protein interactions and that determine the coupling profile of a given receptor. For example, the i1 loop of the formyl peptide receptor (Amatruda et al. 1995) and the cholecystokinin CCK_A receptor (Wu et al. 1997), the i2 loop of the V_{1A} vasopressin receptor (Liu and Wess 1996), and the i3 loop of the endothelin ET_B receptor (Takagi et al. 1995) have been demonstrated to participate in G-protein activation. Taking advantage of chimeric GPCRs designed between structurally related receptor subtypes that are clearly distinguishable with regard to their signaling abilities, studies on muscarinic (Blüml et al. 1994, Blin et al. 1995) and vasopressin receptors (Erlenbach and Wess, 1998) disclosed the importance of several distinct residues for selective G-protein recognition.

The established view of the importance of the intracellular loops in G-protein interaction and specificity is challenged by recent studies with the TSHR and LHR. Large deletions or alanine replacement of most amino acid residues in the i3 loops did not abolish signal transduction, excluding a substantial participation of the i3 loop in G-protein recognition at least in glycoprotein hormone receptors (Wonerow et al. 1998, Schulz et al. 1999, Schulz et al. 2000a). In the m3 muscarinic receptor, a segment of 112 amino acids (central portion of the i3 loop) can be deleted without loss of receptor function (Schöneberg et al. 1995). Similarly, systematic reduction of the length of the i3 loop in the tachykinin NK-1 receptor revealed that most of the loop sequence can be substituted or even deleted without affecting ligand affinity or signal transduction (Nielsen et al. 1998). Further support for the notion that most of the i3 loop sequences are dispensable for G-protein coupling comes from structural comparison of GPCR loop sequences showing that especially the i3 loop varies extremely in length and that there is no obvious sequence homology between GPCRs of similar coupling profiles. Studies with receptor peptides that are able to activate G proteins directly indicate that the cytoplasmic extensions of the TMDs probably provide the surface for G-protein interaction rather than the loops themselves (Abell and Segaloff 1997).

The ability of a GPCR to couple to more than one G-protein subfamily can be conceived as a loss of specificity due to the absence of inhibitory determinants or as a gain of specific contact sites within the receptor molecule. Two general mechanisms explaining multiple coupling events have been suggested – a 'parallel' and a 'sequential' G-protein activation model. In the 'parallel' model the receptor can adopt two or more active conformations (R^*, R^{**}) in an equilibrium. In the 'sequential' G-protein activation model one active conformation (R^*) initiates the signal transduction cascade which modifies the receptor protein (e.g. phosphorylation). The modified receptor can now adopt the second active conformation (R^{**}). Functional and mutational analyses of GPCRs interacting with more than one G-protein family provide evidence for both concepts. Lefkowitz and colleagues offered experimental evidence for a sequential G-protein activation mechanism (Daaka et al. 1997). Agonist-induced activation of the β_2 adrenergic receptor results in cAMP formation via the G_s/adenylyl cyclase pathway followed by cAMP-dependent protein kinase-mediated receptor phosphorylation. Receptor phosphorylation represents a crucial molecular switch mechanism to allow for G_i-mediated ERK activation. These finding are of interest since several primarily G_s-coupled receptors are also capable of activating G_i. Similar results were obtained with the prostacyclin receptor which is primarily coupled to G_s. Following cAMP-dependent protein kinase A activation and receptor phosphorylation at position Ser357 the prostacyclin receptor couples additionally to G_i and G_q (Lawler et al. 2001).

In support of a parallel model of G-protein activation it has been demonstrated that point mutations can selectively abolish receptor coupling to one G-protein subfamily (Surprenant et al. 1992, Gilchrist et al. 1996, Biebermann et al. 1998, Fuchs et al. 2001). On the other hand, the coupling profile of a GPCR can be extended by mutational changes. Concomitantly with G_s activation, the LHR mediates fairly modest agonist-induced phosphoinositide breakdown via G_i recruitment. It was observed that several LHR mutations at the very N-terminal end of TMD6 profoundly enhanced agonist-induced IP accumulation, most likely via $G_{q/11}$ activation (Schulz et al. 1999). These findings are consistent with the concept that GPCRs can exist in at least two distinct active conformations, R* and R**, which differ in their G-protein coupling pattern. Based on this notion, it is also conceivable that different agonists can stabilize distinct ternary complexes. One may speculate that such pathway-selective agonists may represent the protagonists of a new class of therapeutic agents.

3.2
Receptor Domains Interacting with Additional Targets of GPCR Signalling

The four major classes of G-protein α subunits act on well known cellular targets mainly defining the cellular response like changes in the levels of cAMP, IPs, or cations. Upon receptor-mediated GDP release and GTP binding to the α subunit, the α subunit and ßγ subunits dissociate. In addition to the α subunit, free ßγ subunits are modulators of an increasing variety of cellular proteins such as phospholipases, adenylyl cyclases, channels, and kinases (Hamm 1998). More recently, co-precipitation studies, expression cloning strategies and yeast two-hybrid screening approaches have uncovered an ever growing number of targets and co-factors for GPCR signaling (Hall et al. 1999). Most contact sites for co-factor interaction are located in the receptor C termini. For example, the endocytotic sorting of the ß₂ adrenergic receptor is regulated by postsynaptic density/disc-large/ZO1 (PDZ)-

domain-mediated protein interaction. Phosphorylation of a serine by GPCR kinase-5 (GRK-5) within the receptor C terminus enables binding of the phosphoprotein EBP50, thus linking the receptor to the actin cytoskeleton (Cao et al. 1999). Several other PDZ-domain-containing proteins like the Na$^+$/H$^+$ exchanger regulatory factor (NHERF) (Hall et al. 1998), MUPP1 (Ullmer et al. 1998), a subunit of the eukaryotic initiation factor 2B (Klein et al. 1997), and cortactin-binding protein 1 (Zitzer et al. 1999) which interact with a PDZ (S/TXV) motif of GPCRs have been identified. A NHERF-binding site (DS/TxL) in the carboxy-terminal tail of ß$_2$ adrenoceptor becomes available after ligand binding. The carboxy-terminal PDZ domain on NHERF directs its association with the receptor, thereby releasing NHE from the negative regulatory effect of NHERF. It was recently shown that the recruitment of nitric-oxide synthases following stimulation of the serotonin 5-HT$_{2B}$ receptor is mediated via the PDZ motif at the receptor C terminus (Manivet et al. 2000). In contrast, the B$_2$ bradykinin receptor appears to inhibit the endothelial nitric-oxide synthase by direct interaction (Ju et al. 1998).

A mechanism whereby different cells and tissues can regulate cell membrane expression and functional characteristics of GPCRs has recently been disclosed by the discovery of receptor-activity-modifying proteins (RAMPs) (McLatchie et al. 1998). Calcitonin-gene-related peptide (CGRP) and adrenomedullin are related peptides which interact with the same GPCR, the calcitonin-receptor-like receptor (CRLR). Depending on which RAMP ise expressed in a given cell, the CRLR will generate the CGRP and adrenomedullin receptors when associated with RAMP1 and RAMP2, respectively (McLatchie et al. 1998). RAMPs also control glycosylation of the CRLR and transport of the receptor protein to the cell membrane. There are no further examples for an involvement of RAMPs in modifying the function of other GPCR.

A recently identified ER-membrane-associated protein, DRiP78, binds to a conserved motif (FxxxFxxxF) within the D$_1$ dopamine receptor C terminus which functions as an ER-export signal (Bermak et al. 2001). The metabotropic glutamate receptors mGluR$_{1A}$ and mGluR$_5$ contain a Homer ligand binding sequence in their cytoplasmic C terminus which directs the association with the EVH-like domain of Homer proteins. Subsequently, self-multimerization of Homer protein links the mGluRs to IP$_3$ receptors regulating the agonist-independent activity of mGluRs (Tu et al. 1998, Xiao et al. 1998, Ango et al. 2001). Association between the C-terminal portion of the D$_5$ dopamine receptor and the i2 loop of the GABA$_A$ receptor γ$_2$ subunit is likely to be responsible for the functional interaction of these two functionally and structurally divergent receptors (Liu et al. 2000).

Using a yeast two-hybrid approach, p38JAB1, a protein initially identified as cJun co-factor, binds to the C terminus of premature LHR and promotes its degradation (Li et al. 2000). A functional interaction with the human homolog of the Shk1 kinase-binding protein from yeast has been demonstrated for the somatostatin type 1 receptor (Schwarzler et al. 2000). Several catecholamine receptors contain Src homology (SH) 2 and 3 domain binding motifs within the i3 loop enabling the receptor to interact with a variety of SH3 domain-containing proteins (Oldenhof et al. 1998). For example, the SH3 domain of the SH3p4 protein associates with the polyproline motif in the β_1 adrenergic receptor (Tang et al. 1999). The physiological relevance of these many identified interactions remains to be determined.

4
General Aspects of GPCR Dysfunction

The direct and indirect control of an extraordinary variety of physiological functions by GPCRs increase the likelihood that GPCRs are also of pathophysiological relevance. In principle, three different scenarios are conceivable. *First*, the receptor is functionally altered by mutations of the receptor gene or genes necessary for secondary structural features, e.g. posttranslational modification or receptor folding. Mutationally induced GPCR dysfunction can cause gain or loss of physiological function. Generally, such disorders are subdivided into acquired and hereditary malfunctions. Gain-of-function mutations are frequently found in adenomas, but germ-line transmission is rare because of the disease-limited life-time or reproductive ability. Currently, this situation is changing because of progress made in diagnosis and therapy, for example in congenital hyperthyroidism (Grüters et al. 1999). Loss-of-function mutations are associated almost exclusively with hereditary transmission. With a few exceptions which are discussed later, the pathological consequence of loss-of-function mutations become apparent only when both alleles are affected by homozygous or compound heterozygous alterations. Genomic analysis of mutant mice and gene-targeting disruption techniques guided many studies to disease-causing GPCR mutations in man (Lin et al. 1993, Stein et al. 1994, Hosoda et al. 1994). *Second*, a given GPCR is required for the effect of a pathogenic agent (e.g. co-receptors for HIV entry). For example, certain chemokine receptors play a role in HIV pathogenesis, as demonstrated by the occurrence of a CCR_5 deletion mutant, which is frequently found in the Caucasian population and confers a strong, although incomplete protection to homozygotes (Samson et al. 1996, Liu et al. 1996). In contrast, disease-modifying mutations were described for $CX3CR_1$, resulting in a more rapidly progressing form of AIDS (Faure et al. 2000). Which of the 15 known HIV co-receptors are important *in vivo* is poorly defined. And *third*, the pathogenic agent induces the expression of an endogenous GPCR or carries a GPCR in its own genome to cause pathologic effects. The pathophysiologic role of a viral GPCR has been demonstrated for the herpes virus type 8 participating in Kaposi's sarcoma development. It should also be noted that ectopic GPCR expression and permanent receptor stimulation has been implicated in tumor induction and growth (Dhanasekaran et al. 1995, Ferris et al. 1997). In many endocrine tumors, an inappropriate receptor cell surface expression is the cause of significant hormone liberation causing secondary pathological effects (de Herder et al. 1994).

4.1
Diseases Caused by Mutations in GPCRs

Naturally occurring gain-of-function mutations are responsible for only a few human diseases (Table 1). Most mutations are described for the glycoprotein hormone receptors, LHR and TSHR. It is assumed that more than 80% of all toxic or autonomous adenomas of the thyroid are caused by activating mutations in the TSHR (Parma et al. 1997, Tonacchera et al. 1998). Constitutive activation of the G_s/adenylyl cyclase pathway is mainly responsible for the phenotypes found in Jansens's metaphyseal chondrodysplasia, familiar and acquired hyperthyroidism, male-limited precocious puberty, and hypoparathyroidism.

Permanent activation of pathways involving phospholipase C has the potential to promote carcinogenesis as shown for chronic agonist stimulation of G_q-coupling receptors (Allen et al. 1991, Gutkind et al. 1991). Distinct mutations in the TSHR and the LHR can constitutively activate phospholipase C, in addition to the G_s/adenylyl cyclase system, and promote the development of thyroid carcinoma and Leydig-cell tumors, respectively (Russo et al. 1995, Liu et al. 1999). Pathogenicity of viral GPCRs is also associated with constitutive receptor activity. The KSHV-GPCR, closely related to chemokine and interleukin receptors, was found to agonist-independently activate the $G_{q/11}$/PLC signal transduction pathway (Arvanitakis et al. 1997). The KSHV-GPCR contains a VRY sequence instead of the highly conserved DRY motif in the i2 loop (see 2.3). The chemokine receptor $CXCR_2$ is closely related to KSHV-GPCR. The mutant D138V in which the Asp of the DRY motif was replaced by Val exhibited transforming potential similar to the KSHV-GPCR (Burger et al. 1999). Smoothened, a GPCR of the frizzled fam-

Table 1. Diseases caused by mutations in GPCRs

GPCR	Gain-of-function disease	Reference
PTH/PTH-related peptide receptor	Jansen's metaphyseal chondrodysplasia	Schipani et al. 1995
TSHR	Hyperthyroidism, thyroid carcinoma	Parma et al. 1993, Russo et al. 1995
LHR	male-limited precocious puberty, seminoma, Leydig-cell tumor	Shenker et al. 1993, Martin et al. 1998, Liu et al. 1999
Ca^{2+}-sensing receptor	dominant and sporadic hypoparathyroidism	Baron et al. 1996
rhodopsin	night blindness, retinitis pigmentosa	Rao et al. 1994, Robinson et al. 1992
smoothened	sporadic basal-cell carcinoma	Xie et al. 1998
KSHV-GPCR	Kaposi's sarcoma	Arvanitakis et al. 1997

GPCR	Loss-of-function disease	Reference
endothelin B receptor	Hirschsprung's disease	Puffenberger et al. 1994
melanocortin type 1 receptor	UV-induced skin damage	Valverde et al. 1995
melanocortin type 2 receptor	ACTH resistance syndrome	Clark et al. 1993
melanocortin type 4 receptor	dominant and recessive obesity	Yeo et al. 1998, Vaisse et al. 1998, Farooqi et al. 2000
FSHR	ovarian dysplasia, amenorrhea, secondary amenorrhea	Aittomäki et al. 1995, Beau et al. 1998
GnRH receptor	hypogonadotropic hypogonadism	de Roux et al. 1997, Layman et al. 1998
Ca^{2+}-sensing receptor	hyperparathyroidism	Pollak et al. 1993
PTH/PTH-related peptide receptor	Blomstrand chondrodysplasia	Zhang et al. 1998, Jobert et al. 1998
LHR	pseudohermaphroditism, hypospadias	Kremer et al. 1995, Misrahi et al. 1997
TSHR	hypothyroidsm, thyroid hypoplasia	Biebermann et al. 1997
thromboxane A_2 receptor	bleeding disorder	Hirata et al. 1994
$P2Y_{12}$ receptor	bleeding disorder	Hollopeter et al. 2001
hypocretin (orexin) receptor type 2	narcolepsy	Lin et al. 1999
AVPR2	nephrogenic diabetes insipidus (NDI), partial NDI	Rosenthal et al. 1992, Sadeghi et al. 1997a
rhodopsin	retinitis pigmentosa	Dryja et al. 1990
opsin	color blindness	Weitz et al. 1992
retinal GPCR (RGR)	retinitis pigmentosa	Morimura et al. 1999
glucagon receptor	non-insulin-dependent diabetes mellitus	Hager et al. 1995
TRH receptor	isolated central hypothyroidism	Collu et al. 1997
GHRH receptor	Dwarfism	Wajnrajch et al. 1996

ily, utilizes hedgehog and probably also G-protein-dependent signaling pathways (DeCamp et al. 2000, Liu et al. 2001), and activating mutations in this receptor can lead to basal-cell carcinoma (Xie et al. 1998).

The chance of mutational inactivation of receptor proteins is naturally higher than the mutational generation of a constitutively active GPCR. A large number of structural alterations has been described including amino acid substitutions, truncations by nonsense or frameshift mutations, insertions, rearrangement as well as small and large deletions. This diversity of structural modifications, all leading to the same result characterized by impairment or total loss of receptor function, is reflected by more than twenty human diseases known to be caused by inactivating mutations (Table 1). Most diseases display a recessive inheritance, but in some cases, such as dominant obesity (Yeo et al. 1998), the phenotype becomes apparent if only one allele is altered. Besides complete phenotypes, partial loss-of-function or heterozygosity can cause variations in the phenotypical appearance as found for secondary amenorrhea (Beau et al. 1998), hypospadias (Misrahi et al. 1997), partial NDI (Sadeghi et al. 1997), fertile eunuch variant of idiopathic hypogonadotropic hypogonadism (Pitteloud et al. 2001), and hyperthyrotropinemia (Sunthornthepvarakui et al. 1995).

To date, a large number of variants (polymorphisms) in GPCR genes has been identified (for review see Rana et al. 2001). At a polymorphic locus the rarer allele must occur with a frequency greater than 1% in the population. The availability of a reference sequence of the human genome provides the basis for studying the nature of sequence variation, particularly single nucleotide polymorphisms (SNPs), in human populations. SNPs occur at a frequency of approximately 0.5–1 SNP/kb throughout the genome when the sequence of individuals is compared (Mullikin et al. 2000, The genome international sequencing consortium 2001, Venter et al. 2001). SNP typing is a powerful tool for genetic analysis because sequence variants are responsible for the genetic component of individuality, disease susceptibility, and drug response. The latter point will have an important impact on drug design, therapeutic regimes and side effects.

Some SNPs in GPCRs were associated with distinct phenotypes or diseases and functionally characterized *in vitro* (Table 2). Most interesting, a common variant of the ß$_1$ adrenergic receptor (G389R) displays increased constitutive activity, however, no prominent phenotype was observed (Mason et al. 1999). No clearly related phenotype was found for the naturally occurring D$_4$ dopamine receptor deficiency (Nothen et al. 1994). This is in agreement with D$_4$ dopamine receptor-lacking mice displaying locomotor supersensitivity to ethanol, cocaine, and methamphetamine but no obvious phenotype under normal conditions (Rubinstein et al. 1997).

Table 2. Selected mutations/polymorphisms found in GPCRs that were associated with/without distinct human phenotypes

GPCR	Mutation/Polymorphism	Functional consequence / association with	Reference
glucagon receptor	G40S	n.d. / essential hypertension	Brand et al. 1999
calcitonin receptor	silent mutation	n.d. / reduced fracture risk in post-menopausal woman	Taboulet et al. 1998
α_{1A} adrenergic receptor	R492C	no major functional consequences / no obvious phenotype	Shibata et al. 1996
α_{2A} adrenergic receptor	N251K	gain of agonist function in vitro / no phenotype	Small et al. 2000
α_{2B} adrenergic receptor	in-frame deletion	decreased receptor desensitization / reduced metabolic rate	Heinonen et al. 1999
α_{2C} adrenergic receptor	in-frame deletion	n.d. / no obvious phenotype	Feng et al. 2001
β_1 adrenergic receptor	G389R	constitutive activity / no obvious phenotype	Mason et al. 1999
β_2 adrenergic receptor	R16G	enhanced agonist-induced down-regulation / nocturnal asthma	Turki et al. 1995
	Q27E	decreased agonist-induced down-regulation / no obvious phenotype	Green et al. 1994
	T164I	decreased agonist-induced efficacy / no obvious phenotype	Brodde et al. 2001
β_3 adrenergic receptor	W64R	n.d. / glucose intolerance, fatty liver, obesity	Shima et al. 1998
5-HT_{1A} receptor	G22S	attenuated agonist-induced down-regulation / no obvious phenotype	Rotondo et al. 1997
5-HT_{1B} receptor	F124C	increased drug potency / no obvious phenotype	Kiel et al. 2000
5-HT_{2A} receptor	H452Y	no major functional consequences / no obvious phenotype	Arranz et al. 1998
	silent mutation	n.d. / schizophrenia	Williams et al. 1996
5-HT_{2C} receptor	C23S	decreased agonist affinity / no obvious phenotype	Lappalainen et al. 1995
5-HT_{5A} receptor	P15S	n.d. / schizophrenia	Iwata et al. 2001
AT_{1A} receptor	silent mutation	n.d. / hypertension	Van Geel et al. 2000
D_2 dopamine receptor	V154I (heterozygous)	n.d. / myoclonus dystonia	Klein et al. 1999
	P310S, S311C	decreased G_i-coupling efficacy / no obvious phenotype	Cravchik et al. 1996
D_4 dopamine receptor	V194G	decreased agonist affinity / low weight, no axilly or pubic hair	Liu et al. 1996
	deletion (homozygous)	n.d. / acousticus neurinoma, obesity, autonomic nervous system disturbances	Nothen et al. 1994
D_5 dopamine receptor	C335X (heterozygous)	n.d. / no obvious phenotype	Sobell et al. 1995
melatonin 1b receptor	G24E, R54W, L66F,	n.d. / no obvious phenotype	Ebisawa et al. 1999
	A157V (heterozygous)	n.d. / no obvious phenotype	Ebisawa et al. 2000
CCR2	V64I	no major functional consequences / delay in the progression to AIDS	Smith et al. 1997
CX3CR	V249I, T280M	n.d. / progressed to AIDS more rapidly	Faure et al. 2000
CCK_B receptor	E288K	increased drug efficacy / no obvious phenotype	Kopin et al. 1997
μ-opioid receptor	N40D	increased agonist affinity / no obvious phenotype	Bond et al. 1998
P2Y_2	R334C	no major functional consequences / no obvious phenotype	Janssens et al. 1999
melanocortin type 2 receptor	K6T, V8II	n.d. / no obvious phenotype	Hani et al. 2001
melanocortin type 5 receptor	F209L	no major functional consequences / no obvious phenotype	Hatta et al. 2001

n.d. not determined

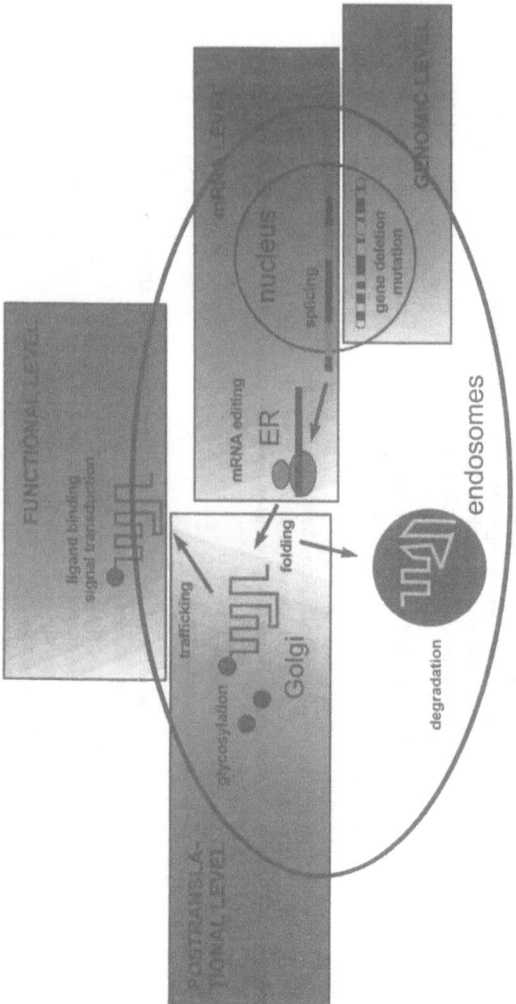

Fig. 4. Molecular mechanisms influencing proper GPCR function. Mutations within GPCR genes can affect receptor function at different levels. At the genomic level, partial or complete gene deletion can interfere with gene transcription. Small genomic alterations can lead to nonsense-mediated mRNA decay, splicing errors and changes in RNA editing (mRNA level). Mutation-induced structural changes in the receptor protein can interfere with proper receptor folding and trafficking (posttranslational level), thus promoting protein degradation in endosomes. Finally, mutations can also abolish receptor signaling by disturbance of the ligand binding domain or G-protein interaction sites (functional level)

Normally, most of the functionally relevant GPCR mutations which do not cause pathologic changes in man are apparently compensated by other mechanisms. However, GPCR mutations can become clinically relevant when challenged (diseases, intoxication). It was shown for a naturally occurring human CCK_B receptor variant (E288K) that the maximal level of receptor-mediated second messenger signaling achieved by synthetic compounds (drug efficacy) is markedly increased compared with values obtained with the wild-type receptor (Kopin et al. 1997). It is likely that completion of the human genome project will lead to a rapid progress in the detection of polymorphisms within distinct GPCR genes.

4.2
Mechanisms of GPCR Alteration

Improper receptor function due to structural alterations can manifest itself at different stages of the GPCR 'life cycle' (Fig. 4). The following section will focus on the molecular mechanisms affecting proper receptor function. Since naturally occurring mutations in the rhodopsin, AVPR2 and the glycoprotein hormone receptors have been studied in great detail, these receptors will serve as model systems to delineate more general principles.

4.2.1
Genomic Level

Like in other hereditary disorders, most diseases caused by GPCR dysfunction are based on genomic alterations with functional effects at different supragenomic levels. Distinct types of genomic alterations can be differentiated – missense mutations, nonsense mutations, small deletions or insertions (in-frame, frameshift) and large or complex deletions. However, the presence of an apparent phenotype depends on the structural impact of the mutation on receptor function. In X-linked myotubular myopathy (Laporte et al. 2000) and retinoblastoma (Lohmann 1999), most point mutations lead to truncations of the myotubularin and RB1 gene product, respectively. In contrast, over 850 different mutations were identified in the cystic fibrosis transmembrane conductance regulator (CFTR) gene coding for a chloride conductance protein, but 70% of all autosomal recessive cystic fibrosis cases are caused by a single triplet deletion (ΔF508) (Zielenski 2000). To date, more than 170 different mutations have been identified within the AVPR2 gene responsible for NDI (Table 3). Out of all clinically relevant mutations detected in GPCRs about 75% are missense mutations. In about 25% of all

Table 3. Frequency of mutations found in selected GPCRs that cause distinct human phenotypes

Except for the AVPR2 and the endothelin B receptor mutation count include inactivating and activating mutations. The following sources were used: The Human Gene Mutation Database (http://archive.uwcm.ac.uk/uwcm/mg/hgmd/), CASRdb database (http://data.mch.mcgill.ca/casrdb/), TSH-receptor database (http://www.uni-leipzig.de/~innere/TSH/), Diabetes insipidus database (http://www.medcon.mcgill.ca/nephros/) and the available literature (http://www.ncbi.nlm.nih.gov/entrez/)

Mutation	AVPR2	Ca^{2+}-sensing receptor	endothelin B receptor	MC-4 receptor	rhodopsin	TSH receptor
Total mutations	172	59	18	23	106	49
	% of total	% of total	% of total	% of total	% of total	% of total
Missense	51.2 (88)	86.4 (51)	66.7 (12)	78.3 (18)	82.1 (87)	85.7 (42)
Nonsense	10.5 (18)	3.4 (2)	11.1 (2)	4.3 (1)	4.7 (5)	4.1 (2)
Small insertion (ins.)	7.6 (13)	5.1 (3)	5.6 (1)	13.0 (3)	0.9 (1)	–
Small deletion (del.)	23.2 (40)	–	5.6 (1)	4.3 (1)	7.5 (8)	6.1 (3)
Small del./ins.	–	1.7 (1)	–	–	–	2.0 (1)
Large/complex ins./del.	5.8 (10)	1.7 (1)	5.6 (1)	–	0.9 (1)	–
Splice site	1.7 (3)	1.7 (1)	5.6 (1)	–	3.8 (4)	2.0 (1)

cases, the mutation resulted in a complete loss or a truncation of the receptor protein due to nonsense mutation, small deletions or insertions and large deletions. Figure 5 depicts all AVPR2 missense mutations and the genomic localization of two large AVPR2 gene deletions. As a consequence of loss-of-function mutations in the AVPR2, the renal response to arginine vasopressin (AVP) is impaired resulting in clinical characteristics of NDI which include hypernatremic dehydration, polyuria, polydipsia, fever, and constipation.

Inactivating mutations in the TSHR have been shown to cause congenital hypothyroidism with variable degree of severity. Besides euthyroid TSH resistance (Sunthornthepvarakui et al. 1995), moderate and severe congenital hypothyroidism caused by compound heterozygosity for inactivating mutations were described in patients with thyroid hypoplasia (Clifton-Bligh et al. 1997, Biebermann et al. 1997).

This large variety of mutations in a member of the GPCR family described above is restricted only to a loss-of-function phenotype. In contrast, gain-of-receptor function is only induced when residues important for receptor quiescence are substituted by distinct amino acids. Therefore, the frequency of activating mutations is lower when compared with inactivating mutations. However, the pathologic consequences caused by constitutive receptor activity are often clinically more apparent, and the underlying genetic defects can usually be identified more easily due to dominant heredity. Additionally, the ability to compensate for somatic mutations leading to constitutive receptor activity appears to be limited.

In the TSHR, both inactivating and activating mutations have been identified (Grüters et al. 1999). Somatic gain-of-function mutations of the TSHR were found in autonomous thyroid adenomas (Parma et al. 1993) and more recently in thyroid insular carcinoma (Russo et al. 1996). Several studies estimated the number of thyroid adenomas caused by activating mutations in the TSHR up to 80% (Parma et al. 1997). However, congenital hyperthyroidism is a rare disease which is mostly caused by transplacental passage of maternal TSHR-stimulating autoantibodies in Graves' disease. Sporadic cases of congenital non-autoimmune hyperthyroidism (Kopp et al. 1995, De Roux et al. 1996) and a few familiar cases of non-autoimmune hyperthyroidism with childhood or later onset (Duprez et al. 1994, Tonacchera et al. 1996, Grüters et al. 1998) were found to be caused by TSHR mutations leading to constitutive activation of the receptor. Activating mutations are mostly located in the TMDs, preferentially in TMD6 and the juxtamembrane portion of the i3 loop and TMD6. Only one amino acid position located within the large extracellular domain (S281) was identified in the TSHR vulnerable for

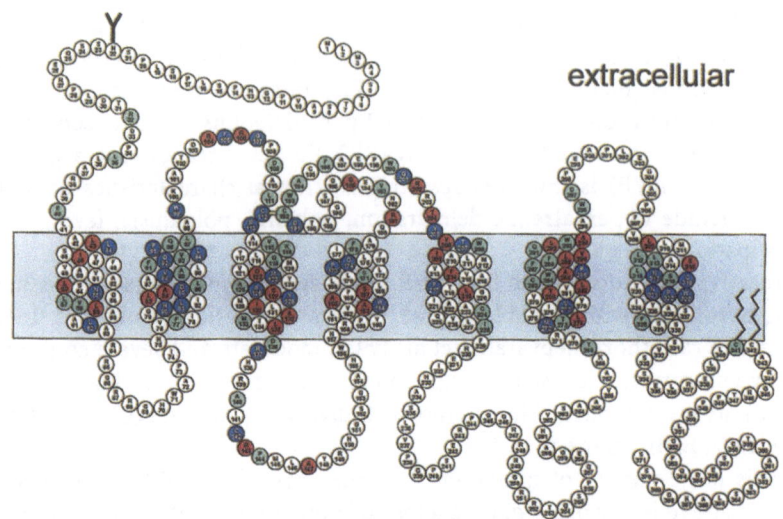

extracellular

○ conserved residue within mammalian vasopressin/oxytocin receptors
● missense mutation and conserved residue within mammalian
  vasopressin/oxytocin receptors
● missense mutation

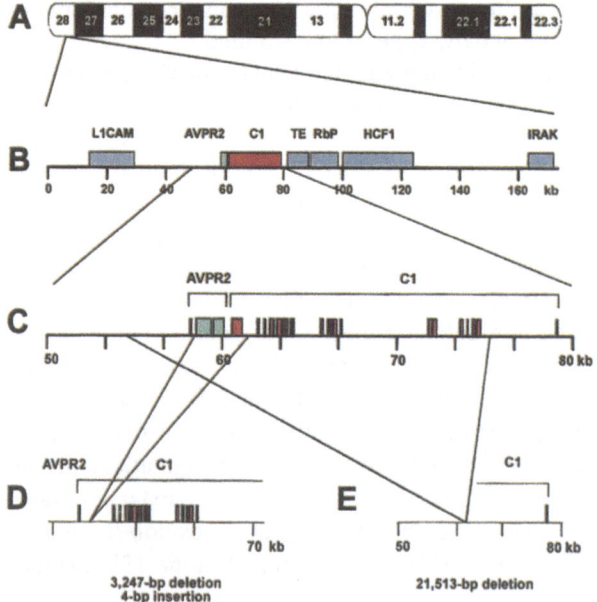

activating mutations (Grüters et al. 1998). Most mutations found resulted in amino acid substitutions, but in two cases in-frame deletions were identified as a cause for hyperthyroidism (Parma et al. 1997, Wonerow et al. 1998).

Such site-specific preferences for gain-of-function mutations are determined functionally. This holds partially true also for inactivating mutations (no inactivating substitution mutations have been reported in the very C terminus of any GPCR), but mutation hotspots are probably based on the presence of vulnerable elements (e.g. CpG sites) within GPCR genes. The mutational mechanisms are multiform. Methylation of DNA is an epigenetic modification that can play an important role in the control of gene expression in mammalian cells. The enzyme involved in this process is DNA methyltransferase, which catalyzes the transfer of a methyl group from S-adenosyl-methionine to cytosine residues to form 5-methylcytosine (5MeC), a modified base that is found mostly at CpG sites in the genome. Many types of DNA damage (oxidative lesions, alkylation of bases, abasic sites, photodimers, etc.) interfere with the ability of mammalian cell DNA to be methylated at CpG dinucleotides by DNA-methyltransferases (DNA-MTases). This can result in altered distribution patterns of 5-methylcytosine (5MeC) residues at CpG sites. CpG sites in DNA represent mutational hotspots, with both the presence of 5MeC in DNA and the catalytic activity of DNA-MTases being intrinsically mutagenic. Many recurrent mutations within the AVPR2

Fig. 5A–E. Model of the AVPR2 and location of missense mutations and large deletions found in patients with X-linked nephrogenic diabetes insipidus. About 22% of all amino acid residues are fully conserved among mammalian vasopressin/oxytocin receptors (green and blue; upper panel). Over 170 different mutations have been identified in the AVPR2 gene of NDI patients. About 50% out of all missense were found at positions which are conserved (blue). Other positions of missense found in NDI patients are shown in red. PCR screening of the Xq28 region of two NDI patients revealed submicroscopic deletions (detailed Xq28 map: lower panels A-C). In one case, the breakpoints were located within the first intron of the AVPR2 gene and the 3'-breakpoint in the last intron (intron 21) of the C1 gene (D). In a second NDI patient (E), the 5'-breakpoint was located within the 5'-untranslated region of the AVPR2 gene and the 3'-breakpoint within the first intron sequence of the C1 gene (panels C, E). To identify the exact positions of the deletion breakpoints, suitable primers pairs were used to amplify interstitial genomic sequence (Schöneberg et al. 1999, Schulz and Schöneberg 2000c). The bars in panel C, D, and E mark the exons of the AVPR2 gene (blue) and the C1 gene (red). Locus abbreviations: L1CAM, neuronal cell adhesion molecule 1, AVPR2, arginine-vasopressin V2 receptor, C1, rhoGAP C1, TE, N-acetyl transferase related protein, RbP, renin-binding protein, HCF1, host cell factor 1, IRAK, interleukin-1 receptor-associated kinase. The nucleotide positions refer to the numbering of the original sequence submission of the Xq28-L1CAM locus (GenBank accession number: U52112)

gene such as D85N, V88M, R106C, R113W, R137H, R181C, R202C, and R337X are probably caused by the hypermutability at such CpG dinucleotides.

Large deletions or rearrangements of chromosomes usually results in fetal abortion and complex syndromes (Tharapel et al. 1993, Ferrero et al. 1997). Submicroscopic alterations have been made responsible for more distinct phenotypes such as mental retardation (Fries et al. 1993), Emery-Dreifuss muscular dystrophy (Small et al. 1997), monoamine oxidase deficiency (Sims et al. 1989), developmental disorders (Gedeon et al. 1995) and myotubular myopathy (Hu et al. 1996). There are only a few reports in which genes encoding for GPCRs are included in microdeletions (Restagno et al. 1993, Laue et al. 1996, Gromoll et al. 2000). To date, only eight submicroscopic deletions or rearrangements have been found in patients with NDI (Bichet et al. 1994, van Lieburget et al. 1995, Jinnouchi et al. 1996, Schöneberg et al. 1999, Arthus et al. 2000, Schulz and Schöneberg 2000c). However, an exact identification of the breakpoints is lacking in most studies but is of considerable interest since the microdeletion can include other functionally important genes. Analysis of the AVPR2 gene of a boy with classical symptoms of NDI revealed a 21.5-kb deletion within Xq28 (Schöneberg et al. 1999). The upstream breakpoint was found within the 5'-untranslated region of the AVPR2 gene and the 3'-breakpoint was localized within the first intron sequence of the Cl gene (Fig. 5). Based on amino acid sequence similarity and functional data, the Cl gene encodes a protein (ARHGAP4) which is thought to function as a Rho GTPase-activating protein (rhoGAP) (Nagase et al. 1995, Tribioli et al. 1996). In a second NDI patient, the 5'-breakpoint is located within the first intron of the AVPR2 gene, and the 3'-breakpoint includes almost half of intron 21, the coding sequence of the C-terminal 77 amino acid residues and the complete 3'-untranslated region (3'-UTR) of the Cl gene (see Fig. 5). Repetitive genomic sequences are often found in close proximity to deletion breakpoints, and identification can help to clarify the mechanism by which the deletion was produced. Since repetitive DNA is abundant in the human genome, it is commonly suggested that microdeletions arise through mispairing of large duplicated sequences (Chen et al. 1997). For example, Alu repeats were found to be involved in homologous and non-homologous recombination events in α°-thalassaemia (Harteveld et al. 1997). Analyses of the breakpoint regions for homologous sequences in both NDI patients (see above) revealed a lack of sequence similarities suggesting a non-homologous recombination event or chromosomal breaks and re-junction as cause for these interstitial deletions. In the absence of any information about the physiological relevance and specific functions of this specific rhoGAP, a thorough clinical and laboratory investigation was initiated. However, both patients did not reveal any major abnormalities besides clearly defined NDI symptoms caused by deletion of the AVPR2 gene.

Phenotype diversity as found in diseases caused by activating and inactivating mutations in GPCRs mainly depends on the location of the alteration and the structural changes made by the mutation. However, other mechanisms such as mosaicism, skewed X-chromosome inactivation and genomic imprinting have to be considered when phenotype diversity is caused by the same mutation. Very early mutation in the patient's genome can result a mosaicism. Only a few cases of mutational mosaicisms have been reported in GPCR-related diseases (Pasel at al. 2000, Biebermann et al. 2001). The occurrence of an inactivating mutation on one allele is usually compensated by the non-mutated second allele (recessive heredity). X-chromosomally located single-copy genes often do not follow this paradigm. In some cases heterozygous occurrence of inactivating mutations in an X-chromosomally located single-copy gene can cause the same disorder as found in hemizygous males (Moses et al. 1995, Nomura et al. 1997). In female embryogenesis, X-chromosome inactivation occurs early at about the 32 to 64 cell stage, and the number of progenitor cells for individual tissues is presumed to be small (Heard et al. 1997). Once an X chromosome has been selected for inactivation, it is irreversible, and the same X chromosome is inactivated in all descendants of that cell. The random process of X-chromosome inactivation results in a normal distribution of skewing among females. The incidence of extreme skewing (90:10) in peripheral blood of 162 normal neonates was 2% (Busque et al. 1996). Skewed X-chromosome inactivation with a partial phenotype of NDI has been observed in female carries of inactivating AVPR2 mutations (Nomura et al. 1997, Arthus et al. 2000, Schulz und Schöneberg 2000c).

An epigenetic imprinting mechanism that is based on a gamete-specific methylation imprint restricts expression of a subset of mammalian genes to one parental chromosome. The most impressive example of genomic imprinting in GPCR signal transduction was described for the G_s α-subunit (Weinstein and Yu 1999). Albright hereditary osteodystrophy is caused by heterozygous inactivating mutations of the gene encoding the G_s α-subunit. When the mutation is inherited from the mother, the offspring will develop pseudohypoparathyroidism type Ia including both hormone resistance and somatic abnormalities. When the mutation is derived from the father, children will have normal parathyroid hormone responses while exhibiting the somatic features of so-called pseudopseudohypoparathyroidism. The *mas* protooncogene encodes a mitogenic GPCR and provides another example for an imprinting phenomenon. It was demonstrated in mice that the maternally inherited allele is transcriptionally repressed in a developmental and tissue-specific manner (Villar and Pedersen 1994).

Heterozygote mutations in the melanocortin-4 receptor gene (MC4-R) were detected which were assumed to lead to the dominant phenotype of extreme obesity. It has been hypothesized that the mutation found in these subjects resulted in a loss of gene function which causes obesity due to haploinsufficiency of the MC4-R gene. However, in contrast to the initial observations (Vaisse et al. 1998, Yeo et al. 1998), moderate obesity mainly applied to males can also be associated with these mutations (Sina et al. 1999). These examples demonstrate that skewed X-chromosome inactivation and genomic imprinting phenomena should be considered if a distinct phenotype caused by inactivating mutations in GPCRs occurs in heterozygote individuals.

4.2.2
Transcriptional Level

Regulation of GPCR transcription and post-transcriptional processes such as splicing and mRNA degradation involves complex cellular mechanisms all fine tuning the receptor resonsiveness (for review see Danner and Lohse 1999). Thus, numerous polymorphisms have been identified within the putative promoter regions, 5' and 3' untranslated (UTR) regions, and introns of GPCR genes (for review see Rana et al. 2001). Many of these polymorphisms in the non-coding regions were associated with human diseases and altered drug responses. However, most studies lack clear experimental evidence for the suggested mechanisms by which the mutation should affect receptor function.

There are only a few reports describing functionally relevant mutations in the donor or acceptor splice sites of GPCRs. In the AVPR2 gene, three NDI-causing mutations in the splice sites of the second intron were identified (Wildin et al. 1994, Wildin et al. 1998, Arthus et al. 2000). Mutations in endothelin B receptor (ET_BR) gene are responsible for congenital aganglionosis (Hirschsprung's disease). In one patient a heterozygous point mutation at the splice donor site of intron 3 was identified leading to premature termination of translation of ET_BR mRNA (Inoue et al. 1998). Besides loss-of-function mutations within the exons of the TSHR, rhodopsin, and the retinal G-protein-coupled receptor (RGR), disease-causing mutations were also found in the intron/exon boundaries of these genes (Macke et al. 1993, Reig et al. 1996, Gagne et al. 1998, Morimura et al. 1999). The impact of mutations affecting receptor mRNA maturation at the posttranscriptional stage is probably underestimated. For example, even 'silent' exon or intron mutations may influence proper mRNA splicing by introducing a new consensus acceptor/donor splice site which, if used, would encode a nonsense or truncated protein. Additionally, it should be taken in consideration (at least theoretically) that functional alteration of the endogenous splicing machinery can indirectly influence normal GPCR function.

All eukaryotes possess the ability to detect and degrade mRNA harboring premature signals for the termination of translation. This ensures that truncated proteins are seldom made, reducing the accumulation of rogue proteins that might be deleterious. Despite the ubiquitous nature of nonsense-mediated mRNA decay and its demonstrated role in the modulation of phenotypes resulting from selected nonsense alleles, very little is known regarding its basic mechanism or the selective pressure for complete evolutionary conservation of this function (Culbertson 1999). For example, 89% of mutations in the *ATM* gene that cause ataxia-telangiectasia (Gilad et al. 1996) lead to premature chain termination, and many of these mutations can trigger nonsense-mediated mRNA decay. As estimated for the AVPR2, about 40% of all mutations will result in a premature stop of receptor protein translation (see 4.2.1). The impact of nonsense-mediated mRNA decay in the pathomechanism of GPCR dysfunction has not been studied yet. Chain-terminating mutations that decrease mRNA abundance by reducing the half-life of mRNA typically behave like loss-of-function alleles. However, nonsense-mediated mRNA decay appears to be not always complete, and some chain-termination mutations confer a dominant phenotype when the RNA-surveillance system is bypassed. Escape from nonsense-mediated mRNA decay can occur by exon skipping due to a nonsense mutation located at a position that affects splice-site selection through changes in RNA structure or a long open reading frame resulting from a mutationally induced frame shift.

The expression of truncated GPCRs has been demonstrated by immunologic and functional methods *in vivo* and *in vitro* (Ridge et al. 1995, Schöneberg et al. 1995, Schöneberg et al. 1996, Heymann and Subramaniam 1997, Grosse et al. 1997, Schöneberg et al. 1998). As demonstrated in stably transfected CHO cells expressing low amounts of truncated AVPR2, the existence of such truncated receptor proteins was not due to saturation of the machinery degrading mutant mRNA (Schöneberg et al. 1997). *In vivo* studies showed that a truncated D_3 dopamine receptor variant (D_{3nf}) is endogenously expressed in monkey and rat cortical neurons (Nimchinsky et al. 1997). Screening of cDNA libraries, expressed sequence tag (EST) databases and genomic analyses revealed that some GPCR genes encode truncated receptors (pseudogenes) as found for a human neuropeptide Y receptor

(Rose et al. 1997), a human 5-HT$_4$-like receptor (Liu et al. 1998), psiGPR32, and psiGPR33 (Marchese et al. 1998).

Dominant-negative effects of truncated GPCRs on wild-type receptor function have been demonstrated for the CCR$_5$ (Benkirane et al. 1997), the AVPR2 (Zhu and Wess 1998), and the GnRH receptor (Grosse et al. 1997). Therefore, it is reasonable to assume that dominant effects caused by prematurely terminated receptor proteins which escaped nonsense-mediated mRNA decay may contribute to the pathomechanism of loss-of-function mutations (see 4.2.3).

4.2.3
Translational and Posttranslational Level

Once a GPCR has been synthesized, the receptor polypeptide adopts a structure which enables the receptor to pass through the ER quality control machinery. Inherited mutations leading to incompletely folded proteins that are retained intracellularly by the ER quality control system are a recurring observation (Cheng et al. 1990, Singh et al. 1997). The basic mechanisms in the process of GPCR folding and trafficking are poorly understood. Despite the lack of detailed information, the folding and trafficking process is recognized as an important part in realizing proper GPCR function. To clearly differentiate between trafficking defects and abolished agonist binding and/or functional coupling, immunologic methods should be applied, in addition to second messenger and radioligand binding assays (Schöneberg et al. 1998).

Numerous investigations have focused on the identification of structural elements within GPCRs necessary to maintain proper receptor folding. The receptor C terminus appears to contain structural information relevant for GPCR folding. Systematic studies with C-terminally truncated or mutated AVPR2s underscored the importance of the C-terminal tail for receptor trafficking (Wenkert et al. 1996, Sadeghi et al. 1997a, Schülein et al. 1998). However, naturally occurring and artificial truncations after TMD7 do virtually not affect the function of other GPCRs, e.g. the gonadotropin-releasing hormone receptor, neurokinin-2 receptor (Alblas et al. 1995), glucagon receptor (Unson et al. 1995), follitropin receptor (Hipkin et al. 1995), LHR (Zhu et al. 1993, Wang et al. 1996), and parathyroid hormone (PTH)/PTH-related peptide receptor (Iida-Klein et al. 1995).

In some cases, it was found that the retention of the mutant protein, but not alteration of its functional properties, is responsible for the disease. For example, it was shown that a mutant CFTR (ΔF508) can form functional cAMP-activated chloride channels when it can escape the ER and be expressed at the cell surface (Zeitlin 2000). Because temperature is known to affect protein folding, Jaquette and Segaloff (1997) examined the effects of reduced temperature on cell surface expression of intracellularly retained mutant LHRs. It was demonstrated that preincubation of the cells for 48 h at 26 °C markedly increased both hCG binding ability and cAMP production, suggesting that decreased temperatures can allow partially misfolded LHRs to fold properly and to be transported to the cell surface.

Impaired insertion of the wild-type receptor into the plasma membrane is a common mechanism underlying the dominant negative effects of co-expressed mutant or truncated receptors. For example, some functionally inactive mutants (N285I, K298X) of the PAF receptor display dominant negative effects on the wild-type receptor due to intracellular retention (Le Gouill et al. 1999). Interestingly, a dominant positive effect, as reflected by constitutive receptor activation, was observed after co-expression of the wild-type and a mutant (D63N) PAF receptor (Le Gouill et al. 1999). It should be noted that dominant negative effects of mutant GPCRs on wild-type receptor function can also be caused by other mechanisms such as titrating G proteins away from the wild-type receptor (Leavitt et al. 1999). Thus, expression of truncated or modified receptor proteins may highlight a novel principle of specific modulation of GPCR function with physiological and pathophysiological relevance. It was demonstrated that a naturally occurring allele coding for a truncated CCR_5 chemokine receptor which functions as a co-receptor for infection by primary M-tropic HIV-1 strains exerts a dominant negative effect on the viral env protein-mediated cell fusion (Samson et al. 1996). It was later shown that the truncated receptor complexes with the wild-type CCR_5 and that this interaction retains CCR_5 in the ER resulting in reduced cell surface expression (Benkirane et al. 1997). In addition, defective intracellular transport due to the formation of misfolded complexes between wild-type and mutant rhodopsin in the ER is held responsible for the dominant effect of one mutant allele in case of retinal degeneration in *Drosophila* (Colley et al. 1995). To understand the pathophysiology of retinitis pigmentosa caused by mutations in the human rhodopsin gene leading to truncated receptor proteins, heterozygote individuals carrying nonsense mutations (Q64X, Q344X) or an intron 4 splice site mutation were studied. Although the three mutations interfered with normal rod cell function, there was allele specificity for the pattern of retinal dysfunction (Jacobson et al. 1994). One may speculate that the observed differences in the phenotype are due to variations in nonsense-mediated mRNA decay. Interestingly, there is one reported case of autosomal dominant retinitis pigmentosa in which the normal termination codon is replaced by a Glu codon. Examination of the genomic sequence revealed that the next termination codon lies 153 bp downstream of the natural stop codon. Termination of translation at this point would add additional 51 amino acid residues to the C terminus of rhodopsin (Bessant et al. 1999).

An other example comes from mutant MC4 receptors that has been found in patients with familiar obesity (see Table 1). The heterozygous occurrence of missense mutations as well as truncating mutations in the MC4 receptor are the cause of the clinical phenotype (Vaisse et al. 1998, Yeo et al. 1998). One may speculate that dominant negative effects of the wild type receptor function contributes to the molecular pathomechanism.

Posttranslational formation of one or more extracellular disulfide bonds is essential for receptor trafficking and high affinity ligand binding (see 1.2.1). Disease-causing mutations of the highly conserved disulfide bond connecting the e1 and the e2 loops have been identified in the AVPR2 (Bichet et al. 1994), rhodopsin (al-Maghtheh et al. 1993, Fuchs et al. 1994) and the ACTH receptor (Naville et al. 1996).

Many of the mutations in the e1 and e2 loops found in patients with NDI and retinitis pigmentosa are characterized by substitutions of various amino acid residues to Cys residues (Sung et al. 1991, Fujiwara et al. 1995, Schöneberg et al. 1998). For example, five such missense mutations, R106C, R181C, G185C, R202C, and Y205C, were identified in the AVPR2 gene (Bichet 1998). These findings together with a phylogenetic sequence comparison (see Fig. 5) suggest that the mutational introduction of only one new Cys residue in the extracellular loops causes altered receptor function. One Cys residue in the e1 loop (Cys 112) and two Cys residues in the e2 loop (Cys192 and Cys195) are available to form a disulfide bond in the wild-type AVPR2. The presence of an additional Cys residue in the extracellular domain caused by mutations may offer alternatives in disulfide bond formation. A recent study by Schülein et al. (2000) offered an additional explanation for the mechanism of AVPR2 dysfunction caused by Cys substitutions. It was shown that an additional C195A mutation results in a functional rescue of the mutant AVPR2s (Schülein et al. 2000). Since C195 is not conserved in mammalians AVPR2s and is not required for wild-type receptor function (Schulz et al. 2000b), these data suggest a mechanism in which C195 forms a new disulfide bond with a mutationally introduced Cys residue, interfering with high affinity binding. Such a mechanism may also account for Cys mutations in rhodopsin leading to an uneven number of Cys residues in the extracellular loops (Schülein et al. 2000).

The polypeptide chain of most GPCRs is posttranslationally modified by N-glycosylation, palmitoylation, and phosphorylation. Clinically relevant mutations of such potential posttranslational modification sites are rare, and only a few examples were published. Substitution mutations altering potential glycosylation sites in rhodopsin were described in families with retinitis pigmentosa (Bunge et al. 1993, Sullivan et al. 1993). A polymorphismus leading to an alteration of a potential N-glycosylation site of the μ-opioid receptor (N40D) is present in almost 10% of the population. This receptor variant displays a higher affinity for ß-endorphin as compared with the wild type receptor (Bond et al. 1998). These findings underline the functional importance of GPCR glycosylation *in vivo*.

Most rhodopsin-like GPCRs display consensus sites for lipid modifications at Cys residues conserved within the C-terminal receptor tail. There are no reports highlighting the importance of this specific post-translational modification *in vivo*. Similarly, no clinical phenotype as a functional consequence of mutations affecting potential phosphorylation consensus sites has been reported yet. Increased receptor phosphorylation has been demonstrated for constitutively active GPCR (Ren et al. 1993, Westphal et al. 1995). Since cross-desensitization of other GPCRs within a given cell is a common phenomenon, one can speculate that the functions of other receptors are also influenced by constitutive receptor activation.

4.2.4
Ligand Binding and Signal Transduction

Once a mutated GPCR has passed through the ER folding control machinery and reached the cell surface, the mutation can still interfere with ligand binding and/or G-protein coupling. There are numerous clinically relevant examples for GPCR mutations showing that loss of receptor function is mainly caused by reduced affinity for the natural ligand (Biebermann et al. 1997, Schöneberg et al. 1998, de Roux et al. 1999, Albertazzi et al. 2000, Morello et al. 2001). For example, an NDI-causing AVPR2 mutation (F105V) which is located in the e1 loop is properly delivered to the cell surface and displays an unchanged maximum cAMP response. Impaired ligand binding was reflected by a concentration-response curve that was shifted towards higher vasopressin concentrations (Pasel et al. 2000). In principle, alteration of the agonist binding pocket can occur either through direct or indirect mechanisms. The first case involves positions that directly participate in ligand interaction. Indirect disturbance of the ligand binding site can occur if residues are mutated that are involved in stabilizing the ligand binding pocket. A clear separation of direct from indirect effects is often difficult.

Isolated defects of receptor/G-protein signaling without effects on receptor trafficking and/or ligand binding are extremely rare. One example comes from Hirschsprung's disease caused by mutations in the ET_BR gene. A mutant ET_BR (W276C) exhibits wild-type binding affinity but interferes with efficient coupling to G_q without affecting $G_{i/o}$-coupling (Imamura et al. 2000). These data suggest that Hirschsprung's disease is mainly caused by defects in the G_q signaling pathway. Controversially, two other ET_BR mutants (G57S and R319W) bound endothelin-1 normally and induced calcium transients in the same way as wild type, but did not inhibit adenylyl cyclase (Fuchs et al. 2001). This may indicate that an additional non-G-protein-mediated signaling pathway is involved in the pathophysiology of ET_BR dysfunction.

Analysis of the biochemical properties of mutant AVPR2s causing a partial NDI phenotype revealed that cell surface expression of D85N (see Fig. 5) is not altered (Sadeghi et al. 1997b). A 50-fold increase in the EC_{50} value and a decreased maximum cAMP response, but almost unchanged affinity, indicate a substantial decrease in the coupling efficiency of D85N to G_s. However, mutation-induced impairment of receptor signaling is usually accompanied by additional defects in receptor trafficking and ligand binding.

Changes in signal transduction are more prominent when the mutation induces agonist-independent receptor activation. Some activating mutations were naturally found (see Table 1), but most constitutively active GPCRs were artificially generated and tested in *in vitro* systems. The i3 loop and TMD6 of GPCRs have been identified as regions that are crucial for receptor/G-protein interactions. Assuming an equal frequency of germline mutations in all GPCR genes, it is astonishing that only a few activating mutations have been identified *in vivo*. Many constitutively active GPCRs may often result in an early letal phenotype. An other reason for the low natural occurrence is that constitutively active receptors are less stable *in vivo*, and, therefore, a clinically relevant phenotype may not always become apparent as shown for a constitutively active ß$_2$ adrenergic receptor in transgenic mice. However, treatment of the transgenic mice with various adrenergic receptor ligands resulted in receptor overexpression and an increase in cardiac functions by stabilizing the mutant receptor protein (Samama et al. 1997).

As stated above, most GPCRs have the capability of coupling to more than one G protein. Distinct receptor/G-protein pathways can be differentially influenced by activating mutations. Selective constitutive activation of only one signal transduction pathway was shown for adrenergic receptors (Perez et al. 1996, Zuscik et al. 1998). Mutational activation of the LHR is the cause of male-limited precocious puberty due constitutive activation of the G_s/adenylyl cyclase pathway (Shenker et al. 1993). In addition to the G_s/adenylyl cyclase system, the activated LHR is known to stimulate G_i and phospholipase C. Depending on the precise nature of the mutation, constitutive LHR activation is observed in the latter signaling pathway which probably accounts for the malign transformation of Leydig cells (Liu et al. 1999). These findings further support the idea of multiple and distinct activation states in GPCRs (see 3.1).

In principle, naturally occurring GPCR mutations may also result in an increase in ligand affinity and/or signal transduction efficiency. As a consequence, lower concentrations (or even basal levels) of the endogenous ligand are sufficient to mediate a maximal physiological response. Such mutant receptors should cause similar pathophysiological effects as found for activating mutations if the dysfunction is not compensated by other mechanisms such as a decrease in agonist concentration or receptor cell surface expression. A pathophysiologically relevant example for this scenario remains to be identified.

5
Therapeutic Strategies to Restore Altered GPCR Function

5.1
General Considerations

To date, the therapy of diseases caused by GPCR dysfunction focuses mainly on treating clinical symptoms. It is beyond the scope of this review to address all therapeutic strategies used to alleviate the symptoms caused by GPCR dysfunction. This section focuses primarily on new strategies which may become therapeutically useful in the future. The most desirable strategy to treat inherited disorders in man is the site-specific reversion of the mutation by restoring the normal nucleotide sequence. However, a major obstacle to successful gene therapy is the relative inefficiency of current targeting strategies in mammalian cells. Gene targeting may be accomplished by homologous recombination and mismatch correction of DNA heteroduplexes. Homologous recombination, initially considered to be of limited use for gene therapy because of its low frequency in mammalian cells, has recently emerged as a potential strategy. Different approaches have been used including small or large homologous DNA fragments, adenovirus-associated viral vectors, and RNA/DNA chimeric oligonucleotides. One promising approach is based on recent observations of recombinogenic activity of specifically designed chimeric RNA/DNA oligonucleotides (Cole-Strauss et al. 1996, Rando et al. 2000, Alexeev et al. 2000). However, the gene repair efficiency remains low and has not been tested in clinically relevant settings.

Treatment of diseases caused by GPCR dysfunction may not always require genetic approaches. For example, loss-of-function mutations in the TSHR have been demonstrated to cause some familial forms of athyreosis with an absolute thyroxine (T_4) deficiency resulting in developmental and mental retardation (Biebermann et al. 1997). Early discovery of the genetic defect by neonatal screening and rapid clinical management are important for patients, as these patients will have normal fertility. The goals of treatment are to raise the serum T_4 as rapidly as possible into the normal range to maintain normal growth and development (Grüters et al. 1999). Simple hormone substitution is also the treatment of choice in hypogonadotropic hypogonadism (Zitzmann and Nieschlag 2000)

5.2
Ligand Binding-Directed Strategies

It has been shown for a large number of inactivating mutations in GPCRs that alteration of the agonist binding site reduces ligand affinity but leaves maximal coupling efficacy unchanged. Therefore, patients with such inactivating mutations may benefit from high agonist doses or from administration of an agonist displaying a higher affinity to the mutant receptor than the natural ligand. The usefulness of this principle has been demonstrated in dogs suffering from NDI because of a low affinity AVPR2. Prolonged treatment with high doses of an AVPR2 specific agonist, 1-deamino [D-Arg8] vasopressin (dDAVP), rendered the NDI-affected dogs near normal in terms of water intake and urine osmolality (Luzius et al. 1992). Recombinant TSH 'superagonists' with increased receptor affinity and efficacy were designed for the TSHR by combining evolutionary considerations, sequence comparisons and homology modeling (Grossmann et al. 1998). Success in treatment of TSHR and any other GPCR dysfunction with high affinity agonists in man is still lacking.

In the traditional model, the unliganded receptor was assumed to be quiescent, but differences in basal second messenger levels depending on the receptor type and expression levels have questioned this theory. The simplest model addressing the latter phenomenon suggests that receptors exist in an equilibrium between two conformations, an inactive stage R and an active stage R* (see 3.1). Agonists are thought to alter the position of the equilibrium by stabilizing R* (see Fig. 3). Antagonists do not preferentially stabilize either state of the receptor but compete for agonist or inverse agonist binding. Inverse agonists stabilize the R state, thus leading to decreased basal receptor activity. In many cases high levels of (ectopically) expressed GPCRs can produce substantial constitutive activity, and an inverse agonist would be preferable for eliminating GPCR-induced cellular responses rather than an antagonist (Milligan et al. 1995, Milligan and Bond 1997). Because GPCRs are thought to exist in an equilibrium between R and R*, an increase

in the total receptor number will also increase the total number with the R*
conformation which may become pathophysiologically relevant (Bond et al.
1995). In thyreotoxicosis, the number of cardiac ß adrenergic receptors is
increased causing an increase in heart-beat frequency. A combination of
thyreostatic drugs and ß adrenergic blockers with negative intrinsic activity
are used in the treatment of this life-threatening condition.

One may speculate that inverse agonists would become the number one
choice in the treatment of diseases caused by activating mutations. To date,
positive results in the treatment of patients suffering form diseases caused
by activating GPCR mutations with an inverse agonist have not been reported. There are, however, a few experimental examples suggesting a
benefit of this strategy. For instance, mutations of Lys at position 296 (K296E
and K296M) found in certain patients with autosomal dominant retinitis
pigmentosa leads to constitutive activity of rhodopsin. It was shown that the
11-*cis* C19 retinylamine is an effective inhibitor of the naturally occurring
constitutively active mutant rhodopsins (Yang et al. 1997). Constitutive
activity of the viral receptor KSHV-GPCR is an important factor in Kaposi's
sarcoma pathogenesis. It has been demonstrated recently that the human
interferon (IFN)-γ-inducible protein 10 (HuIP-10), a CXC chemokine, and
the stromal cell-derived factor-1 α specifically inhibit signaling of KSHV-
GPCR and KSHV-GPCR-induced proliferation of NIH 3T3 cells (Geras-
Raaka et al. 1998a, Rosenkilde et al. 1999). These data may promote the development of non-peptide inverse agonist drugs for the treatment of Kaposi's sarcoma.

Unfortunately, the use of an inverse agonist may not always represent a
suitable therapeutic strategy. Although a compound may behave as an inverse agonist of the wild-type receptor, it may become a potent and efficient
agonist of a constitutively active receptor mutant. Several examples of mutation-induced changes in GPCR pharmacological properties have been
reported, these include cholecystokinin (Beinborn et al. 1998) and bradykinin (Marie et al. 1999) mutant receptors.

5.3
GPCR Expression- and Downregulation-Directed Strategies

Suppression of receptor mRNA transcription as well as induction of receptor
downregulation and degradation may also represent a suitable concept for
silencing constitutively active GPCRs. As already mentioned above, Kaposi's
sarcoma-associated herpes virus, which is consistently present in tissues of
patients with Kaposi's sarcoma and primary effusion lymphomas, contains a
gene that encodes a constitutively active GPCR (KSHV-GPCR). It was shown

that co-expression of GPCR-specific kinases and activation of protein kinase C inhibit constitutive KSHV-GPCR signaling (Geras-Raaka et al. 1998b).

In heart failure, compensatory mechanisms involve an increased activity of the sympathetic nervous system mainly due to elevated norepinephrine plasma levels ('quasi' constitutive activation). This increase in agonism can lead to ß adrenergic receptor desensitization and, therefore, in loss of receptor function. ß Adrenergic receptor desensitization is mediated through enhanced activity of the ß adrenergic receptor kinase (ßARK1) in ischemic and failing myocardium. Adenoviral-mediated gene delivery of an inhibitor of ßARK1, ßARKct, preserves normal ß adrenergic receptor function presumably by inhibiting desensitization through endogenous β_1 adrenergic receptors (White et al. 2000).

On the other hand, constitutively elevated levels of GPCR agonists can be useful in the treatment of diseases. In congestive heart failure, high systemic levels of AVP result in vasoconstriction and reduced cardiac contractility mediated by the V_1 vasopressin receptor. The AVPR2 is physiologically expressed in the kidney but not in the myocardium. Ectopic expression of a recombinant AVPR2 in the myocardium which promotes activation of the G_s/adenylyl cyclase system (like the β_1 adrenergic receptor) results in a positive inotropic effect in heart failure, mediated by the high endogenous AVP levels (Laugwitz et al. 1999). In contrast to the findings with adrenergic receptors, AVPR2 expression does not undergo significant downregulation.

Another promising novel approach takes advantage of specifically engineered GPCRs referred to as RASSLs (receptor activated solely by a synthetic ligand). Specifically, the κ opioid receptor, which couples to $G_{i/o}$, was modified to respond exclusively to synthetic small molecule agonists and not to their natural agonists (Coward et al. 1998). RASSL-transgenic mice were used to control physiologic events depending on activation of the $G_{i/o}$ signal transduction pathway in a tissue specific manner (Redfern et al. 1999). Recently, Liggett and co-workers generated a β_2 adrenergic receptor which was activated only by a non-biogenic amine agonist which itself failed to activate the wild-type receptor (Small et al. 2001). RASSL may have an implication in specific targeting of cells or tissues for diagnostic and therapeutic purposes.

5.4
Receptor Folding-Directed Strategies

To date, methods aimed at correcting genetic defects at the genomic level are of limited effectiveness. This requires a search for therapeutic alternatives. Based on findings that GPCRs are composed of multiple folding units (see 2.4), it was demonstrated that mutant AVPR2s containing clinically relevant mutations in the C-terminal third of the receptor protein can be functionally rescued by co-expression with a non-mutated C-terminal AVPR2 fragment (Schöneberg et al. 1996). Since such polypeptides are expected to interact only with those mutant receptors from which they are derived, this approach would offer the great advantage that the fragments exert their pharmacological effects only in tissues or cells in which the mutant receptors are physiologically expressed. Therefore, embedding of a tissue-specific promoter in the expression cassette of the therapeutic delivery system is not essentially required. To test the potential therapeutic usefulness of this co-expression strategy, cell lines stably expressing low levels of functionally inactive mutant AVPR2s were infected with a recombinant adenovirus carrying a AVPR2 gene fragment encoding the C-terminal third of the receptor protein. Adenovirus-mediated expression of receptor fragments resulted in cell-specific molecular correction of functional receptor defects, indicating that this approach may lead to novel strategies in the treatment of diseases caused by inactivating GPCR mutations (Schöneberg et al. 1997). These findings are probably relevant not only for GPCRs but also for polytopic integral membrane proteins in general. For example, it has been demonstrated that many proteins containing multiple membrane-spanning domains can be assembled from two or more protein fragments. These proteins include sodium channels (Stühmer et al. 1989), the a-factor transporter in yeast (Berkower and Michaelis 1991), adenylyl cyclases (Tang et al. 1995), GLUT1 glucose transporter (Cope et al. 1994), and lactose permease (Bibi and Kaback 1990, Zen et al. 1994).

Therapeutic approaches aiming at the modulation of chaperone function were pioneered by studies with CFTR. Mutations in the CFTR gene reduce the probability of the mutant protein to dissociate from molecular chaperones and largely prevent protein maturation through the secretory pathway to the plasma membrane. These mutant CFTR molecules are rapidly degraded by cytoplasmic proteasomes (Kopito 1999). Treatment of cells expressing the $\Delta F508$ mutant with a number of low molecular weight compounds such as glycerol, dimethylsulfoxide and trimethylamine oxide, all known to stabilize proteins in their native conformation ('chemical chaperones'), results in the correct processing of the mutant CFTR protein and its delivery to the plasma membrane, thus restoring CFTR function (Brown et al. 1996, Sato et al. 1996). Trafficking-deficient aquaporin-2 proteins, as found in patients with recessive NDI, were also redistributed normally after treatment with 'chemical chaperones' (Tamarappoo and Verkman 1998).

Treatment of cells expressing the H_2 histamine receptor with the inverse agonists cimetidine and ranitidine led to an increase of receptor number probably by stabilizing the receptor structure (Smit et al. 1996). Stabilizing effects and receptor up-regulation by inverse agonist treatment were also observed with a constitutively active TRH receptor (Heinflink et al. 1995) and the α_{1B}-adrenergic receptor (Lee et al. 1997). Recently, this approach has been successfully used to functionally rescue clinically relevant AVPR2 mutants by pretreatment of transfected cells with the nonpeptidic AVPR2 antagonists, SR121463A and VPA-985 (Morello et al. 2000). Therefore, the effects of temperature (see 4.2.3), chemical chaperones and structure-stabilizing ligands on the intracellular processing of mutant GPCRs suggest that strategies designed to promote protein folding/stability *in vivo* may eventually lead to the development of novel therapies for diseases caused by GPCR dysfunction.

5.5
Modulation of Receptor Function by Receptor-Derived Polypeptides

The exact nature of GPCR/G-protein interaction sites is currently unknown and may vary between different GPCRs and G proteins. It is assumed that not only the intracellular loops but also the cytoplasmic sides of the TMDs participate in GPCR/G-protein coupling. Indeed, peptides derived from the i3 loop/TMD6 junction can activate the G_s/adenylyl cyclase system (Abell and Segaloff 1997, Abell et al. 1998) or inhibit adenylyl cyclase via G_i proteins (Varrault et al. 1994). Several other studies using expression plasmids coding for cytoplasmic receptor fragments demonstrated impaired receptor/G-protein coupling (Ulloa-Aguirre et al. 1998, Thompson et al. 1998). Recently, the usefulness of dominant negative mutations introduced into the G_s α-subunit was demonstrated as a specific tool for dissection of G_s-mediated signals in cultured cells (Iiri et al. 1999). Since these approaches appear to be specific for a single signaling cascade but lack receptor specificity, such methods will probably have a limited therapeutic relevance.

In accord with the multiple folding unit model of GPCRs, specific inhibition of GPCR function can be achieved by co-expression or administration of single TMDs. It was shown for the ß$_2$ adrenergic receptor, the D$_1$ dopamine receptor, and the CCK$_A$ receptor that TMD6-derived peptides were able to concentration-dependently inhibit receptor signaling (Hebert et al. 1996, George et al. 1998, Tarasova et al. 1999). A potential therapeutic usefulness of this approach was shown for CXCR$_4$ and CCR$_5$ chemokine receptors which serve as co-receptors for HIV-1 entry. Peptides derived from chemokine receptor TMDs can serve as potent and specific receptor antagonists and can block HIV-1 replication *in vitro* (Tarasova et al. 1999).

5.6
Animal Models Suitable for Testing New Therapeutic Strategies

Genomic analysis of mutant mice and gene-targeting disruption techniques guided many studies to disease-causing GPCR mutations in man (Lin et al. 1993, Stein et al. 1994, Hosoda et al. 1994). A large number of GPCRs has been targeted in mice to study the phenotype induced by gene disruption (reviewed in Offermanns 2000). About 50% of GPCR-defective mouse strains show a defined phenotype, about 40% have an obvious phenotype after challenging. This data indicate that the number of uncovered phenotypes caused by mutated GPCR genes in man will increase in the near future.

Only a limited number of mouse strains are available carrying phenotype-causing point mutations in GPCR genes opening the possibility to test therapeutic strategies to restore altered GPCR function. Several examples are discussed in the following. The hyt/hyt hypothyroid mouse displays an autosomal recessive, fetal-onset, severe hypothyroidism. The mutants are characterized by retarded growth, infertility, mild anemia, elevated serum cholesterol, very low to undetectable serum thyroxine, and elevated serum thyroid-stimulating hormone. Thyroid glands are in the normal location but are reduced in size and hypoplastic. Mutant mice respond to thyroid hormone therapy by improved growth and fertility (Beamer et al. 1981). Genetic analysis of the hyt/hyt locus at chromosome 12 revealed a single base exchange in the TSHR gene which leads to the replacement of a highly conserved Pro at amino acid position 556 (TMD4) with Leu (Stein et al. 1994). Mutant mice respond to thyroid hormone therapy by improved growth and fertility but alternative approaches besides hormone substitution to rescue thyroid function have not been tested.

The little mouse is a dwarf strain characterized by low levels of growth hormone, pituitary hypoplasia, and an unresponsiveness to treatment with exogenous growth hormone-releasing hormone (GRHR). Genetic mapping and cloning studies have localized this defect to a point mutation in the N-terminal extracellular domain of the GRHR receptor, a member of family 2 GPCRs, where an Asp residue at position 60 is mutated to Gly (Godfrey et al. 1993). Consistent with the dwarf phenotype, the mutant receptor is inactive, and cells expressing the mutant receptor do not accumulate cAMP in response to GHRH.

Mouse coat color genes have long been studied as a paradigm for genetic interactions in development. Piebald-lethal mice exhibit a recessive phenotype identical to that of the ET_BR knockout mice (Hirschsprung's disease). Southern blotting revealed a deletion encompassing the entire ET_BR gene in

Piebald-lethal mice (Hosoda et al. 1994). However, reconstitution of the ET_BR gene by genetic or other approaches has not been attempted so far.

In X-linked NDI more than hundred inactivating mutations have been reported within the AVPR2 gene (Bichet 1998). Current pharmacological treatment strategies of X-linked NDI include the administration of thiazide diuretics along with a reduction in salt intake, frequently in combination with prostaglandin synthesis inhibitors. However, these drugs only lead to a partial reduction in urine production, and their use is often associated with severe side effects including disturbances in electrolyte balance as well as renal and gastrointestinal complications. An animal model of X-linked NDI should allow detailed studies of the pathophysiology of this disease and should greatly facilitate the development of new therapeutic strategies including rescue attempts with receptor fragments. Premature chain-termination mutations have been successfully used in mice to induce loss-of-function of the D_3 dopamine receptor (Accili et al. 1996) and the 5-HT$_{1A}$ receptor (Heisler et al. 1998). Therefore, a nonsense mutation known to cause X-linked NDI in humans (E242X) was introduced into the mouse genome (Yun et al. 2000). AVPR2-deficient male (–/y) mice die within the first week after birth, apparently due to hypernatremia and dehydration caused by the inability of these animals to concentrate urine. The body weight and the urine osmolality levels of the (–/y) mice are significantly reduced. Interestingly, adult female AVPR2 (+/–) mice show clear symptoms of NDI, including reduced urine-concentrating ability, polyuria, and polydipsia. The AVPR2-deficient mice will assist in the development of novel therapeutic strategies including functional reconstitution attempts at the genomic and protein levels.

There are only a few examples in the literature reporting animal phenotypes caused by activating mutations. Activating receptor mutations were found in the mouse melanocortin-1 receptor (E92K, L98P) resulting in dark coat color (Lu et al. 1998). Mice expressing constitutive active receptors in a tissue-specific manner were generated for the PTH/PTHrP receptor (Schipani et al. 1997) and the ß$_2$ adrenergic receptor (Samama et al. 1997) and may serve as model systems for therapeutic approaches in the future.

6
Summary and Future Perspectives in GPCR Research

Sequencing of entire genomes of many higher organisms will be completed in the next few years. This information will allow the identification of new GPCRs which are more distantly related to known receptors and will help to understand the function of human GPCRs as well. Hundreds of 'orphan'

GPCRs have been discovered so far. Defining the role of each GPCR under physiological and pathophysiological circumstances and using this information to control GPCR activity therapeutically represents one of the major challenges for molecular medicine in the coming post-genome era. Recent molecular characterization of cloned protein genes draws attention to alternative splicing and mRNA editing as a source of structural and functional diversity. Tissue-specific splicing has also been reported for several GPCRs and is likely to further increase the number of known GPCR isoforms. GPCR diversity is primarily based on the existence of multiple receptor subtypes due to gene duplication events. The existence of introns in GPCR genes provides the potential for additional diversity by virtue of alternative splicing events which may generate distinct receptor isoforms (Kilpatrick et al. 1999).

A more detailed understanding of the function of GPCRs will be achieved as we identify sequences that control receptor expression. Numerous studies have been undertaken aimed at the identification and functional characterization of GPCR promoter sequences. The major questions that still need to be addressed dealing with the molecular basis of time- and tissue-specific expression as well as cellular inputs necessary for GPCR expression. By comparing corresponding genomic sequences in different species, e.g. man, mouse, chicken, and zebrafish, regions that have been highly conserved during evolution can be identified, many of which reflect conserved functions such as gene regulation. These approaches, in combination with mouse genetics, promise to greatly accelerate our understanding of regulation and expression of GPCR genes.

Undoubtedly, resolving the crystal structure of rhodopsin was a milestone in the history of GPCR research (Palczewski et al. 2000). This structural information provides a solid basis for further experimental structure/function relationship studies and for computer models of other GPCRs. Clearly, the success in crystallizing an integral membrane receptor will encourage projects to increase the resolution of the rhodopsin model and to analyze other GPCRs. Future attempts will aim at resolving the fine structure of different functional states in GPCR activation and the co-crystallization of a GPCR in complex with the G protein or other GPCR-associated proteins. These studies should eventually provide detailed structural information about the molecular architecture of the receptor/transducer interface. It should be noted that direct physical methods such as X-ray crystallography can only provide information of a static structure. There are numerous examples showing distinct differences in the fine structure when NMR derived structures in solution are compared with those obtained with X-ray crystallography (Barbato et al. 1999, Erbel et al. 1999, Yuan et al. 1999). Structural information about the receptor protein in its natural environment obtained from biophysical methods as well as mutagenesis and biochemical approaches is still needed to supplement current GPCR models. The correctness of GPCR models will be measured by the success of computer-aided drug design and the correct prediction of the functional consequences of specific mutations.

Identification of a growing number of receptor-binding proteins suggests alternative signaling mechanisms. Structural determinants within the C terminus and the intracellular loops provide potential interaction sites with other cellular proteins. Such partners may serve as direct signaling target or adapter proteins in the down-stream signaling pathway. Yeast two-hybrid and co-immunoprecipitation experiments are powerful methods for identifying novel proteins that bind to GPCRs. These same techniques, coupled with mutagenesis experiments, have been used to define the regions of interaction between pairs of proteins. Thus, an increasing number of non-G-protein interaction partners has been identified during the past few years (see 3.2, for review see Heuss and Gerber 2000). Detailed analyses of the interface for protein interaction revealed a large set of modules, such as SH domains and PDZ domains mediating the GPCR/protein association. The diversity of relevant interactions is likely to grow steadily, particularly since homologous and heterologous GPCR oligomerization has begun to emerge as a rather general phenomenon. Functional characterization of these alternative signaling pathways and the elucidation of the molecular basis of their interactions will advance our understanding of GPCR function in the future.

Human genome research has made it possible to identify the presence of gene mutations in persons with inherited disorders, who may be carriers of genetic disorders, or who are at risk for future development of inherited diseases. Molecular analysis of inherited disorders, population screening of genetic diseases, as well as studies of the genetic basis of variable drug response involve efficient screening for mutations in multiple DNA samples. Thus, high throughput mutation screening methods are of great importance. Traditional methods for mutation screening like SSCP and heteroduplex analysis lack sensitivity and are difficult to automate. However, recent developments in DNA fragment analysis by capillary electrophoresis and microchip-based technologies have made fully automated mutation screening possible and have dramatically increased the possible sample throughput. This offers the possibility to include promoter and intron sequences into the screening routine at a financially acceptable level. The concept that variations in non-coding regions may affect GPCR function is derived from extensive studies with a non-G-protein-coupled membrane receptor. Familial hypercholesterolemia is not only caused by mutations in the coding sequence of the low density lipoprotein (LDL) receptor gene but also by mutations in the promoter region (Koivisto et al. 1994). Such direct linkage of promoter polymorphisms/ mutations with a human disease phenotype has not yet been demonstrated for GPCRs. Numerous studies have been undertaken to link polymorphisms found in the 5'-untranslated region of adrenergic, serotonin and dopamine receptors to psychiatric disorders, but clear proof is still missing. A recent study provides evidence for linkage between CCR$_5$ promoter polymorphisms and long-term asymptomatic HIV-1 infection (Clegg et al. 2000). Genetic variability of GPCR genes influencing cardiovascular phenotypes in normal persons is likely to be relevant to cardiovascular disease. The potential clinical relevance of polymorphisms found in the 3'-untranslated region of the angiotensin receptor type 1 gene associated with arterial hypertension, aortic stiffness and coronary artery disease is currently under investigation (Castellano et al. 1996). Improved understanding of the complex regulation of GPCR expression will probably uncover clinically relevant alterations in the non-coding regions of GPCR genes.

The prospects for the development of gene therapy treatments for certain diseases have been fuelled by advances in the understanding of the molecular basis of these disorders. Certainly, the number of diseases caused by GPCR malfunction will increase in the future. As for other genetic defects, the primary goal of a therapeutic intervention will be the reversal of the disease-causing mutation at the genomic level. Since this approach is limited by low recombination efficiency, most genetic strategies aim at the supragenomic level such as transfer of complete expression cassettes. The ability to transfer and appropriately express the genes encoding molecules of interest is dependent on the availability of effective gene transfer vectors.

After more than one century since the term 'receptor' has been introduced by Paul Ehrlich and after almost two decades since the first GPCR, rhodopsin, was cloned by Nathans and Hogness (1983), this field of research is still exponentially expanding. Long established views are changing since GPCR dimerization, alternative pathways in GPCRs signaling and their pathomechanistic contributions in human diseases have been discovered. Defining the role of each GPCR and using this information to control GPCR activity therapeutically represents one of the major challenges for molecular medicine in the coming post-genome era.

Acknowledgements. We like to thank Jürgen Wess for suggestions and critically reading the manuscript. We also have to acknowledge many people for discussions and exchange of ideas; our particular thanks go to Annette Grüters, Heike Biebermann, Klaus-Peter Hofmann, and Gerd Krause. The

authors own work was supported by the Deutsche Forschungsgemeinschaft
and the Fonds der Chemischen Industrie.

Abbreviations

AVP	arginine vasopressin
AVPR2	V2 vasopressin receptor
ßARK1	ß adrenergic receptor kinase
BRET	bioluminescence resonance energy transfer
CCK-R	cholecystokinin receptor
CD	circular dichroism
CFTR	cystic fibrosis transmembrane conductance regulator
CGRP	calcitonin-gene-related peptide
CRLR	calcitonin-receptor-like receptor
DNA-Mtases	DNA-methyltransferases
e1-e3 loops	extracellular loops 1-3
ET-R	endothelin receptor
FRET	fluorescence resonance energy transfer
FSHR	follitropin receptor
GIRK	G-protein-regulated inward rectifier K^+ channel
GnRH	gonadotropin-releasing hormone
GPCR	G-protein-coupled receptor
GRK	G-protein-coupled receptor kinase
hCG	human choriogonadotropin
i1-i3 loops	intracellular loops 1-3
IP	inositol phosphate
KSHV-GPCR	Kaposi's sarcoma-associated herpes virus receptor
LHR	lutropin receptor
NDI	X-linked nephrogenic diabetes insipidus
NMR	nuclear magnetic resonance
PAF	platelet-activating factor
PDZ	postsynaptic density/disc-large/ZO1
RAMP	receptor-activity-modifying protein
rhoGAP	Rho GTPase-activating protein
SNP	single nucleotide polymorphism
SSCP	single-strand conformational polymorphism
SSTR	somatostatin receptor
TMD	transmembrane domain
TRH	thyrotropin-releasing hormone
TSHR	thyrotropin receptor

References

AbdAlla S, Zaki E, Lother H, Quitterer U (1999) Involvement of the amino terminus of the B_2 receptor in agonist-induced receptor dimerization. J Biol Chem 274:26079-84

AbdAlla S, Lother H, Quitterer U (2000) AT1-receptor heterodimers show enhanced G-protein activation and altered receptor sequestration. Nature 407:94-8

Abdulaev NG, Ridge KD (1998) Light-induced exposure of the cytoplasmic end of transmembrane helix seven in rhodopsin. Proc Natl Acad Sci USA 95:12854-9

Abell AN, Segaloff DL (1997) Evidence for the direct involvement of transmembrane region 6 of the lutropin/choriogonadotropin receptor in activating G_s. J Biol Chem 272:14586-91

Abell AN, McCormick DJ, Segaloff DL (1998) Certain activating mutations within helix 6 of the human luteinizing hormone receptor may be explained by alterations that allow transmembrane regions to activate G_s. Mol Endocrinol 12:1857-69

Accili D, Fishburn CS, Drago J, Steiner H, Lachowicz JE, Park BH, Gauda EB, Lee EJ, Cool MH, Sibley DR, Gerfen CR, Westphal H, Fuchs S (1996) A targeted mutation of the D_3 dopamine receptor gene is associated with hyperactivity in mice. Proc Natl Acad Sci USA 93:1945-9

Acharya S, Karnik SS (1996) Modulation of GDP release from transducin by the conserved Glu134-Arg135 sequence in rhodopsin. J Biol Chem 271:25406-11

Aittomaki K, Lucena JL, Pakarinen P, Sistonen P, Tapanainen J, Gromoll J, Kaskikari R, Sankila EM, Lehvaslaiho H, Engel AR (1995) Mutation in the follicle-stimulating hormone receptor gene causes hereditary hypergonadotropic ovarian failure. Cell 82:959-68

Albertazzi E, Zanchetta D, Barbier P, Faranda S, Frattini A, Vezzoni P, Procaccio M, Bettinelli A, Guzzi F, Parenti M, Chini B (2000) Nephrogenic diabetes insipidus: functional analysis of new AVPR2 mutations identified in Italian families. J Am Soc Nephrol 11:1033-43

Alblas J, van Etten I, Khanum A, Moolenaar WH (1995) C-terminal truncation of the neurokinin-2 receptor causes enhanced and sustained agonist-induced signaling. Role of receptor phosphorylation in signal attenuation. J Biol Chem 270:8944-51

Alewijnse AE, Timmerman H, Jacobs EH, Smit MJ, Roovers E, Cotecchia S, Leurs R (2000) The effect of mutations in the DRY motif on the constitutive activity and structural instability of the histamine H_2 receptor. Mol Pharmacol 57:890-8

Alexeev V, Igoucheva O, Domashenko A, Cotsarelis G, Yoon K (2000) Localized in vivo genotypic and phenotypic correction of the albino mutation in skin by RNA-DNA oligonucleotide. Nat Biotechnol 18:43-7

Allen LF, Lefkowitz RJ, Caron MG, Cotecchia S (1991) G-protein-coupled receptor genes as protooncogenes: constitutively activating mutation of the α_{1B}-adrenergic receptor enhances mitogenesis and tumorigenicity. Proc Natl Acad Sci USA 88:11354-8

al-Maghtheh M, Gregory C, Inglehearn C, Hardcastle A, Bhattacharya S (1993) Rhodopsin mutations in autosomal dominant retinitis pigmentosa. Hum Mutat 2:249-55

Altenbach C, Yang K, Farrens DL, Farahbakhsh ZT, Khorana HG, Hubbell WL (1996) Structural features and light-dependent changes in the cytoplasmic interhelical E-F loop region of rhodopsin: a site-directed spin-labeling study. Biochemistry 35:12470-8

Amatruda TT, Dragas-Graonic S, Holmes R, Perez HD (1995) Signal transduction by the formyl peptide receptor. Studies using chimeric receptors and site-directed mutagenesis define a novel domain for interaction with G-proteins. J Biol Chem 270:28010-3

Angers S, Salahpour A, Joly E, Hilairet S, Chelsky D, Dennis M, Bouvier M (2000) Detection of ß2-adrenergic receptor dimerization in living cells using bioluminescence resonance energy transfer (BRET). Proc Natl Acad Sci USA 97:3684-9

Ango F, Prezeau L, Muller T, Tu JC, Xiao B, Worley PF, Pin JP, Bockaert J, Fagni L (2001) Agonist-independent activation of metabotropic glutamate receptors by the intracellular protein Homer. Nature 411:962-5

Ardati A, Goetschy V, Gottowick J, Henriot S, Valdenaire O, Deuschle U, Kilpatrick GJ (1999) Human CRF2 a and ß splice variants: pharmacological characterization using radioligand binding and a luciferase gene expression assay. Neuropharmacology 38:441-8

Arora KK, Cheng Z, Catt KJ (1997) Mutations of the conserved DRS motif in the second intracellular loop of the gonadotropin-releasing hormone receptor affect expression, activation, and internalization. Mol Endocrinol 11:1203-12

Arranz MJ, Munro J, Owen MJ, Spurlock G, Sham PC, Zhao J, Kirov G, Collier DA, Kerwin RW (1998) Evidence for association between polymorphisms in the promoter and coding regions of the 5-HT2A receptor gene and response to clozapine. Mol Psychiatry 3:61-6

Arthus MF, Lonergan M, Crumley MJ, Naumova AK, Morin D, De Marco LA, Kaplan BS, Robertson GL, Sasaki S, Morgan K, Bichet DG, Fujiwara TM (2000) Report of 33 novel AVPR2 mutations and analysis of 117 families with X-linked nephrogenic diabetes insipidus. J Am Soc Nephrol 11:1044-54

Arvanitakis L, Geras-Raaka E, Varma A, Gershengorn MC, Cesarman E (1997) Human herpesvirus KSHV encodes a constitutively active G-protein-coupled receptor linked to cell proliferation. Nature 385:347-50

Avissar S, Amitai G, Sokolovsky M (1983) Oligomeric structure of muscarinic receptors is shown by photoaffinity labeling: subunit assembly may explain high- and low-affinity agonist states. Proc Natl Acad Sci USA 80:156-9

Bai J, Bishop JV, Carlson JO, DeMartini JC (1999) Sequence comparison of JSRV with endogenous proviruses: envelope genotypes and a novel ORF with similarity to a G-protein-coupled receptor. Virology 258:333-43

Bai M, Trivedi S, Brown EM (1998) Dimerization of the extracellular calcium-sensing receptor (CaR) on the cell surface of CaR-transfected HEK293 cells. J Biol Chem 273:23605-10

Baker EK, Colley NJ, Zuker CS (1994) The cyclophilin homolog NinaA functions as a chaperone, forming a stable complex in vivo with its protein target rhodopsin. EMBO J 13: 4886-96

Baldwin JM (1994) Structure and function of receptors coupled to G proteins. Curr Opin Cell Biol 6:180-90

Ballesteros JA, Jensen AD, Liapakis G, Rasmussen SG, Shi L, Gether U, Javitch JA (2001) Activation of the ß2-adrenergic receptor involves disruption of an ionic lock between the cytoplasmic ends of transmembrane segments 3 and 6. J Biol Chem 276:29171-7

Barak LS, Oakley RH, Laporte SA, Caron MG (2001) Constitutive arrestin-mediated desensitization of a human vasopressin receptor mutant associated with nephrogenic diabetes insipidus. Proc Natl Acad Sci USA 98:93-8

Barbato G, Cicero DO, Nardi MC, Steinkuhler C, Cortese R, De Francesco R, Bazzo R (1999) The solution structure of the N-terminal proteinase domain of the hepatitis C virus (HCV) NS3 protein provides new insights into its activation and catalytic mechanism. J Mol Biol 289:371-84

Baron J, Winer KK, Yanovski JA, Cunningham AW, Laue L, Zimmerman D, Cutler GB Jr (1996) Mutations in the Ca^{2+}-sensing receptor gene cause autosomal dominant and sporadic hypoparathyroidism. Hum Mol Genet 5:601-6

Beamer WJ, Eicher EM, Maltais LJ, Southard JL (1981) Inherited primary hypothyroidism in mice. Science 212:61-3

Beau I, Touraine P, Meduri G, Gougeon A, Desroches A, Matuchansky C, Milgrom E, Kuttenn F, Misrahi M (1998) A novel phenotype related to partial loss of function mutations of the follicle stimulating hormone receptor. J Clin Invest 102:1352-9

Beck M, Sakmar TP, Siebert F (1998) Spectroscopic evidence for interaction between transmembrane helices 3 and 5 in rhodopsin. Biochemistry 37:7630-39

Beinborn M, Quinn SM, Kopin AS (1998) Minor modifications of a cholecystokinin-B/gastrin receptor non-peptide antagonist confer a broad spectrum of functional properties. J Biol Chem 273:14146-51

Benkirane M, Jin DY, Chun RF, Koup RA, Jeang KT (1997) Mechanism of transdominant inhibition of CCR5-mediated HIV-1 infection by ccr5Δ32. J Biol Chem 272:30603-6

Berkower C, Michaelis S (1991) Mutational analysis of the yeast a-factor transporter STE6, a member of the ATP binding cassette (ABC) protein superfamily. EMBO J 10:3777-85

Bermak JC, Li M, Bullock C, Zhou QY (2001) Regulation of transport of the dopamine D1 receptor by a new membrane-associated ER protein. Nat Cell Biol 3:492-8

Bessant DA, Khaliq S, Hameed A, Anwar K, Payne AM, Mehdi SQ, Bhattacharya SS (1999) Severe autosomal dominant retinitis pigmentosa caused by a novel rhodopsin mutation (Ter349Glu). Mutations in brief no. 208. Online. Hum Mutat 13:83

Bibi E, Kaback HR (1990) In vivo expression of the lacY gene in two segments leads to functional lac permease. Proc Natl Acad Sci USA 87:4325-9

Bichet DG (1998) Nephrogenic diabetes insipidus. Am J Med 105:431-42

Bichet DG, Birnbaumer M, Lonergan M, Arthus MF, Rosenthal W, Goodyer P, Nivet H, Benoit S, Giampietro P, Simonetti S, Fish A, Whitley CB, Jaeger P, Gertner J, New M, DiBona FJ, Kaplan BS, Robertson GL, Hendy GN, Fujiwara TM, Morgan K (1994) Nature and recurrence of AVPR2 mutations in X-linked nephrogenic diabetes insipidus. Am J Human Genetics 55:278-86

Biebermann H, Grüters A, Schöneberg T, Gudermann T (1997) Congenital hypothyroidism caused by mutations in the thyrotropin-receptor gene. N Engl J Med 336:1390-1

Biebermann H, Schöneberg T, Schulz A, Krause G, Grüters A, Schultz G, Gudermann T (1998) A conserved tyrosine residue (Y601) in transmembrane domain 5 of the human thyrotropin receptor serves as a molecular switch to determine G-protein coupling. FASEB J 12:1461-71

Biebermann H, Schöneberg T, Hess C, Germak J, Gudermann T, Grüters A (2001) The first activating TSHR mutation in transmembrane domain 1 identified in a family with non-autoimmune hyperthyroidism. J Clin Endocrinol Metab 86:4429-33

Black DL (1998) Splicing in the inner ear: a familiar tune, but what are the instruments? Neuron 20:165-8

Blanpain C, Lee B, Vakili J, Doranz BJ, Govaerts C, Migeotte I, Sharron M, Dupriez V, Vassart G, Doms RW, Parmentier M (1999) Extracellular cysteines of CCR5 are required for chemokine binding, but dispensable for HIV-1 coreceptor activity. J Biol Chem 274:18902-8

Blin N, Yun J, Wess J (1995) Mapping of single amino acid residues required for selective activation of $G_{q/11}$ by the m3 muscarinic acetylcholine receptor. J Biol Chem 270:17741-8

Blüml K, Mutschler E, Wess J (1994) Identification of an intracellular tyrosine residue critical for muscarinic receptor-mediated stimulation of phosphatidylinositol hydrolysis. J Biol Chem 269:402-5

Bond C, LaForge KS, Tian M, Melia D, Zhang S, Borg L, Gong J, Schluger J, Strong JA, Leal SM, Tischfield JA, Kreek MJ, Yu L (1998) Single-nucleotide polymorphism in the human mu opioid receptor gene alters ß-endorphin binding and activity: possible implications for opiate addiction. Proc Natl Acad Sci USA 95:9608-13

Bond RA, Leff P, Johnson TD, Milano CA, Rockman HA, McMinn TR, Apparsundaram S, Hyek MF, Kenakin TP, Allen LF (1995) Physiological effects of inverse agonists in transgenic mice with myocardial overexpression of the ß₂-adrenoceptor. Nature 374:272-6

Bouvier M, Chidiac P, Hebert TE, Loisel TP, Moffett S, Mouillac B (1995) Dynamic palmitoylation of G-protein-coupled receptors in eukaryotic cells. Methods Enzymol 250:300-14

Brand E, Bankir L, Plouin PF, Soubrier F (1999) Glucagon receptor gene mutation (Gly40Ser) in human essential hypertension: the PEGASE study. Hypertension 34:15-7

Brodde OE, Buscher R, Tellkamp R, Radke J, Dhein S, Insel PA (2001) Blunted cardiac responses to receptor activation in subjects with Thr164Ile ß2-adrenoceptors. Circulation 103:1048-50

Brown CR, Hong-Brown LQ, Biwersi J, Verkman AS, Welch WJ (1996) Chemical chaperones correct the mutant phenotype of the ΔF508 cystic fibrosis transmembrane conductance regulator protein. Cell Stress Chaperones 1:117-25

Bunge S, Wedemann H, David D, Terwilliger DJ, van den Born LI, Aulehla-Scholz C, Samanns C, Horn M, Ott J, Schwinger E (1993) Molecular analysis and genetic mapping of the rhodopsin gene in families with autosomal dominant retinitis pigmentosa. Genomics 17:230-3

Burger M, Burger JA, Hoch RC, Oades Z, Takamori H, Schraufstatter IU (1999) Point mutation causing constitutive signaling of CXCR2 leads to transforming activity similar to Kaposi's sarcoma herpesvirus-G protein-coupled receptor. J Immunol 163:2017-22

Burns CM, Chu H, Rueter SM, Hutchinson LK, Canton H, Sanders-Bush E, Emeson RB (1997) Regulation of serotonin-2C receptor G-protein coupling by RNA editing. Nature 387:303-8

Busque L, Mio R, Mattioli J, Brais E, Blais N, Lalonde Y, Maragh M, Gilliland DG (1996) Nonrandom X-inactivation patterns in normal females: Lyonization ratios vary with age. Blood 88:59-65

Cao TT, Deacon HW, Reczek D, Bretscher A, von Zastrow M (1999) A kinase-regulated PDZ-domain interaction controls endocytic sorting of the ß$_2$-adrenergic receptor. Nature 401:286-90

Castellano M, Muiesan ML, Beschi M, Rizzoni D, Cinelli A, Salvetti M, Pasini G, Porteri E, Bettoni G, Zulli R, Agabiti-Rosei E (1996) Angiotensin II type 1 receptor A/C1166 polymorphism. Relationships with blood pressure and cardiovascular structure. Hypertension 28:1076-80

Cetani F, Tonacchera M, Vassart G (1996) Differential effects of NaCl concentration on the constitutive activity of the thyrotropin and the luteinizing hormone/chorionic gonadotropin receptors. FEBS Lett 378:27-31

Chatterjee TK, Sharma RV, Fisher RA (1996) Molecular cloning of a novel variant of the pituitary adenylate cyclase-activating polypeptide (PACAP) receptor that stimulates calcium influx by activation of L-type calcium channels. J Biol Chem 271:32226-32

Chen, KS, Manian P, Koeuth T, Potocki L, Zhao Q, Chinault AC, Lee CC, Lupski JR (1997) Homologous recombination of a flanking repeat gene cluster is a mechanism for a common contiguous gene deletion syndrome. Nat Genet 17:154-63

Cheng SH, Gregory RJ, Marshall J, Paul S, Souza DW, White GA, O'Riordan CR, Smith AE (1990) Defective intracellular transport and processing of CFTR is the molecular basis of most cystic fibrosis. Cell 63:827-34

Claeysen S, Sebben M, Becamel C, Bockaert J, Dumuis (1999) A Novel brain-specific 5-HT$_4$ receptor splice variants show marked constitutive activity: role of the C-terminal intracellular domain. Mol Pharmacol 55:910-20

Clark AJ, McLoughlin L, Grossman A (1993) Familial glucocorticoid deficiency associated with point mutation in the adrenocorticotropin receptor. Lancet 341:461-2

Clegg AO, Ashton LJ, Biti RA, Badhwar P, Williamson P, Kaldor JM, Stewart GJ (2000) CCR5 promoter polymorphisms, CCR5 59029A and CCR5 59353C, are under represented in HIV-1-infected long-term non-progressors. The Australian Long-Term Non-Progressor Study Group. AIDS 14:103-8

Clifton-Bligh RJ, Gregory JW, Ludgate M, John R, Persani L, Asteria C, Beck-Peccoz P, Chatterjee VK (1997) Two novel mutations in the thyrotropin (TSH) receptor gene in a child with resistance to TSH. J Clin Endocrinol Metab 82:1094-100

Cole-Strauss A, Yoon K, Xiang Y, Byrne BC, Rice MC, Gryn J, Holloman WK, Kmiec EB (1996) Correction of the mutation responsible for sickle cell anemia by an RNA-DNA oligonucleotide. Science 273:1386-9

Colley NJ, Cassill JA, Baker EK, Zuker CS (1995) Defective intracellular transport is the molecular basis of rhodopsin-dependent dominant retinal degeneration. Proc Natl Acad Sci USA 92:3070-4

Collu R, Tang J, Castagne J, Lagace G, Masson N, Huot C, Deal C, Delvin E, Faccenda E, Eidne KA, Van Vliet G (1997) A novel mechanism for isolated central hypothyroidism: inactivating mutations in the thyrotropin-releasing hormone receptor gene. J Clin Endocrinol Metab 82:1561-5

Conn PM, Rogers DC, Stewart JM, Niedel J, Sheffield T (1982) Conversion of a gonadotropin-releasing hormone antagonist to an agonist. Nature 296:653-5

Cook JV, Eidne KA (1997) An intramolecular disulfide bond between conserved extracellular cysteines in the gonadotropin-releasing hormone receptor is essential for binding and activation. Endocrinology 138:2800-6

Cope DL, Holman GD, Baldwin SA, Wolstenholme AJ (1994) Domain assembly of the GLUT1 glucose transporter. Biochem J 300:291-4

Coward P, Wada HG, Falk MS, Chan SD, Meng F, Akil H, Conklin BR (1998) Controlling signaling with a specifically designed G$_i$-coupled receptor. Proc Natl Acad Sci USA 95:352-7

Cravchik A, Sibley DR, Gejman PV (1996) Functional analysis of the human D2 dopamine receptor missense variants. J Biol Chem 271:26013-7

Culbertson MR (1999) RNA surveillance. Unforeseen consequences for gene expression, inherited genetic disorders and cancer. Trends Genet 15:74-80

Daaka Y, Luttrell LM, Lefkowitz RJ (1997) Switching of the coupling of the α_2-adrenergic receptor to different G proteins by protein kinase A. Nature 390:88-91

Danner S, Lohse MJ (1999) Regulation of beta-adrenergic receptor responsiveness modulation of receptor gene expression. Rev Physiol Biochem Pharmacol 136:183-223

de Herder WW, Krenning EP, Malchoff CD, Hofland LJ, Reubi JC, Kwekkeboom DJ, Oei HY, Pols HA, Bruining HA, Nobels FR (1994) Somatostatin receptor scintigraphy: its value in tumor localization in patients with Cushing's syndrome caused by ectopic corticotropin or corticotropin-releasing hormone secretion. Am J Med 96:305-12

de Roux N, Polak M, Couet J, Leger J, Czernichow P, Milgrom E, Misrahi M (1996) A neomutation of the thyroid-stimulating hormone receptor in a severe neonatal hyperthyroidism. J Clin Endocrinol Metab 81:2023-6

de Roux N, Young J, Brailly-Tabard S, Misrahi M, Milgrom E, Schaison G (1999) The same molecular defects of the gonadotropin-releasing hormone receptor determine a variable degree of hypogonadism in affected kindred. J Clin Endocrinol Metab 84:567-72

de Roux N, Young J, Misrahi M, Genet R, Chanson P, Schaison G, Milgrom E (1997) A family with hypogonadotropic hypogonadism and mutations in the gonadotropin-releasing hormone receptor. N Engl J Med 337:1597-602

Debouck C, Metcalf B (2000) The impact of genomics on drug discovery. Annu Rev Pharmacol Toxicol 40:193-207

DeCamp DL, Thompson TM, de Sauvage FJ, Lerner MR (2000) Smoothened activates G$_{\alpha i}$ mediated signaling in frog melanophores. J Biol Chem 275:26322-7

de Lean A, Stadel JM, Lefkowitz RJ (1980) A ternary complex model explains the agonist-specific binding properties of the adenylate cyclase-coupled ß-adrenergic receptor. J Biol Chem 255:7108-17

Dhanasekaran N, Heasley LE, Johnson GL (1995) G protein-coupled receptor systems involved in cell growth and oncogenesis. Endocr Rev 16:259-70

Dryja TP, McGee TL, Reichel E, Hahn LB, Cowley GS, Yandell DW, Sandberg MA, Berson EL (1990) A point mutation of the rhodopsin gene in one form of retinitis pigmentosa. Nature 343:364-6

Duprez L, Parma J, Van Sande J, Allgeier A, Leclere J, Schvartz C, Delisle MJ, Decoulx M, Orgiazzi J, Dumont J, Vassart G (1994) Germline mutations in the thyrotropin receptor gene cause non-autoimmune autosomal dominant hyperthyroidism. Nat Genet 7:396-401

Ebisawa T, Kajimura N, Uchiyama M, Katoh M, Sekimoto M, Watanabe T, Ozeki Y, Ikeda M, Jodoi T, Sugishita M, Iwase T, Kamei Y, Kim K, Shibui K, Kudo Y, Yamada N, Toyoshima R, Okawa M, Takahashi K, Yamauchi T (1999) Alleic variants of human melatonin 1a receptor: function and prevalence in subjects with circadian rhythm sleep disorders. Biochem Biophys Res Commun 262:832-7

Ebisawa T, Uchiyama M, Kajimura N, Kamei Y, Shibui K, Kim K, Kudo Y, Iwase T, Sugishita M, Jodoi T, Ikeda M, Ozeki Y, Watanabe T, Sekimoto M, Katoh M, Yamada N, Toyoshima R, Okawa M, Takahashi K, Yamauchi T (2000) Genetic polymorphisms of human melatonin 1b receptor gene in circadian rhythm sleep disorders and controls. Neurosci Lett 280:29-32

Elling CE, Nielsen SM, Schwartz TW (1995) Conversion of antagonist-binding site to metal-ion site in the tachykinin NK-1 receptor. Nature 374:74-7

Erbel PJ, Karimi-Nejad Y, De Beer T, Boelens R, Kamerling JP, Vliegenthart JF (1999) Solution structure of the a-subunit of human chorionic gonadotropin. Eur J Biochem 260:490-8

Erlenbach I, Wess J (1998) Molecular basis of V_2 vasopressin receptor/G_s coupling selectivity. J Biol Chem 273:26549-58

Farahbakhsh ZT, Ridge KD, Khorana HG, Hubbell WL (1995) Mapping light-dependent structural changes in the cytoplasmic loop connecting helices C and D in rhodopsin: a site-directed spin labeling study. Biochemistry 34:8812-9

Farooqi IS, Yeo GS, Keogh JM, Aminian S, Jebb SA, Butler G, Cheetham T, O'Rahilly S (2000) Dominant and recessive inheritance of morbid obesity associated with melanocortin 4 receptor deficiency. J Clin Invest 106:271-9

Farrens DL, Altenbach C, Yang K, Hubbell WL, Khorana HG (1996) Requirement of rigid-body motion of transmembrane helices for light activation of rhodopsin. Science 274:768-70

Faure S, Meyer L, Costagliola D, Vaneensberghe C, Genin E, Autran B, Delfraissy JF, McDermott DH, Murphy PM, Debre P, Theodorou I, Combadiere C (2000) Rapid progression to AIDS in HIV+ individuals with a structural variant of the chemokine receptor CX3CR1. Science 287:2274-7

Feng J, Zheng J, Gelernter J, Kranzler H, Cook E, Goldman D, Jones IR, Craddock N, Heston LL, Delisi L, Peltonen L, Bennett WP, Sommer SS (2001) An in-frame deletion in the α_{2C} adrenergic receptor is common in African--Americans. Mol Psychiatry 6:168-72

Ferreira PA, Nakayama TA, Pak WL, Travis GH (1996) Cyclophilin-related protein RanBP2 acts as a chaperone for red/green opsin. Nature 383:637-40

Ferrero GB, Gebbia M, Pilia G, Witte D, Peier A, Hopkin RJ, Craigen WJ, Shaffer LG, Schlessinger D, Ballabio A, Casey B (1997) A submicroscopic deletion in Xq26 associated with familial situs ambiguus. Am J Hum Genet 61:395-401

Ferris HA, Carroll RE, Rasenick MM, Benya RV (1997) Constitutive activation of the gastrin-releasing peptide receptor expressed by the nonmalignant human colon epithelial cell line NCM460. J Clin Invest 100:2530-7

Franke RR, Sakmar TP, Graham RM, Khorana HG (1992) Structure and function in rhodopsin. Studies of the interaction between the rhodopsin cytoplasmic domain and transducin. J Biol Chem 267:14767-74

Fries MH, Lebo RV, Schonberg SA, Golabi M, Seltzer WK, Gitelman SE, Golbus MS (1993) Mental retardation locus in Xp21 chromosome microdeletion. Am J Med Genet 46:363-8

Fuchs S, Amiel J, Claudel S, Lyonnet S, Corvol P, Pinet F (2001) Functional characterization of three mutations of the endothelin B receptor gene in patients with Hirschsprung's disease: evidence for selective loss of G_i coupling. Mol Med 7:115-24

Fuchs S, Kranich H, Denton MJ, Zrenner E, Bhattacharya SS, Humphries P, Gal A (1994) Three novel rhodopsin mutations (C110F, L131P, A164V) in patients with autosomal dominant retinitis pigmentosa. Hum Mol Genet 3:1203

Fujiwara TM, Morgan K, Bichet DG (1995) Molecular biology of diabetes insipidus. Annu Rev Med 46:331-43

Gagne N, Parma J, Deal C, Vassart G, Van Vliet G (1998) Apparent congenital athyreosis contrasting with normal plasma thyroglobulin levels and associated with inactivating mutations in the thyrotropin receptor gene: are athyreosis and ectopic thyroid distinct entities? J Clin Endocrinol Metab 83:1771-5

Gales C, Kowalski-Chauvel A, Dufour MN, Seva C, Moroder L, Pradayrol L, Vaysse N, Fourmy D, Silvente-Poirot S (2000) Mutation of Asn-391 within the conserved NPXXY motif of the cholecystokinin B receptor abolishes G_q protein activation without affecting its association with the receptor. J Biol Chem 275:17321-7

Gedeon AK, Meinanen M, Ades LC, Kaariainen H, Gecz J, Baker E, Sutherland GR, Mulley JC (1995) Overlapping submicroscopic deletions in Xq28 in two unrelated boys with developmental disorders: identification of a gene near FRAXE. Am J Hum Genet 56:907-14

George SR, Fan T, Xie Z, Tse R, Tam V, Varghese G, O'Dowd BF (2000) Oligomerization of μ- and κ-opioid receptors. Generation of novel functional properties. J Biol Chem 275:26128-35

George SR, Lee SP, Varghese G, Zeman PR, Seeman P, Ng GY, O'Dowd BF (1998) A transmembrane domain-derived peptide inhibits D_1 dopamine receptor function without affecting receptor oligomerization. J Biol Chem 273:30244-8

Geras-Raaka E, Varma A, Ho H, Clark-Lewis I, Gershengorn MC (1998a) Human interferon-gamma-inducible protein 10 (IP-10) inhibits constitutive signaling of Kaposi's sarcoma-associated herpesvirus G protein-coupled receptor. J Exp Med 188:405-8

Geras-Raaka E, Arvanitakis L, Bais C, Cesarman E, Mesri EA, Gershengorn MC (1998b) Inhibition of constitutive signaling of Kaposi's sarcoma-associated herpesvirus G protein-coupled receptor by protein kinases in mammalian cells in culture. J Exp Med 187:801-6

Gether U, Lin S, Ghanouni P, Ballesteros JA, Weinstein H, Kobilka BK (1997) Agonists induce conformational changes in transmembrane domains III and VI of the ß₂ adrenoceptor. EMBO J 16:6737-47

Gether U, Lin S, Kobilka BK (1995) Fluorescent labeling of purified ß₂ adrenergic receptor. Evidence for ligand-specific conformational changes. J Biol Chem 270:28268-75

Ghanouni P, Steenhuis JJ, Farrens DL, Kobilka BK (2001) Agonist-induced conformational changes in the G-protein-coupling domain of the beta 2 adrenergic receptor. Proc Natl Acad Sci USA 98:5997-6002

Ghanouni P, Schambye H, Seifert R, Lee TW, Rasmussen SG, Gether U, Kobilka BK (2000) The effect of pH on ß₂ adrenoceptor function. Evidence for protonation-dependent activation. J Biol Chem 275:3121-7

Gilad S, Khosravi R, Shkedy D, Uziel T, Ziv Y, Savitsky K, Rotman G, Smith S, Chessa L, Jorgensen TJ, Harnik R, Frydman M, Sanal O, Portnoi S, Goldwicz Z, Jaspers NG, Gatti RA, Lenoir G, Lavin MF, Tatsumi K, Wegner RD, Shiloh Y, Bar-Shira A (1996) Predominance of null mutations in ataxia-telangiectasia. Hum Mol Genet 5:433-9

Gilchrist RL, Ryu KS, Ji I, Ji TH (1996) The luteinizing hormone/chorionic gonadotropin receptor has distinct transmembrane conductors for cAMP and inositol phosphate signals. J Biol Chem 271:19283-7

Glusman G, Yanai I, Rubin I, Lancet D (2001) The complete human olfactory subgenome. Genome Res 11:685-702

Godfrey P, Rahal JO, Beamer WG, Copeland NG, Jenkins NA, Mayo KE (1993) GHRH receptor of little mice contains a missense mutation in the extracellular domain that disrupts receptor function. Nat Genet 4:227-32

Gouldson PR, Snell CR, Bywater RP, Higgs C, Reynolds CA (1998) Domain swapping in G-protein coupled receptor dimers. Protein Eng 11:1181-93

Green SA, Turki J, Innis M, Liggett SB (1994) Amino-terminal polymorphisms of the human beta 2-adrenergic receptor impart distinct agonist-promoted regulatory properties. Biochemistry 33:9414-9

Gromoll J, Eiholzer U, Nieschlag E, Simoni M (2000) Male hypogonadism caused by homozygous deletion of exon 10 of the luteinizing hormone (LH) receptor: differential action of human chorionic gonadotropin and LH. J Clin Endocrinol Metab 85:2281-6

Grosse R, Schöneberg T, Schultz G, Gudermann T (1997) Inhibition of gonadotropin-releasing hormone receptor signaling by expression of a splice variant of the human receptor. Mol Endocrinol 11:1305-18

Grossmann M, Leitolf H, Weintraub BD, Szkudlinski MW (1998) A rational design strategy for protein hormone superagonists. Nat Biotechnol 16:871-5

Grüters A, Krude H, Biebermann H, Liesenkotter KP, Schöneberg T, Gudermann T (1999) Alterations of neonatal thyroid function. Acta Paediatr Suppl 88:17-22

Grüters A, Schöneberg T, Biebermann H, Krude H, Krohn HP, Dralle H, Gudermann T (1998) Severe congenital hyperthyroidism caused by a germ-line neo mutation in the extracellular portion of the thyrotropin receptor. J Clin Endocrinol Metab 83:1431-6

Gudermann T, Schöneberg T, Schultz G (1997) Functional and structural complexity of signal transduction via G-protein-coupled receptors. Annu Rev Neurosci 20:399-427

Guiramand J, Montmayeur JP, Ceraline J, Bhatia M, Borrelli E (1995) Alternative splicing of the dopamine D_2 receptor directs specificity of coupling to G-proteins. J Biol Chem 270:7354-8

Gutkind JS, Novotny EA, Brann MR, Robbins KC (1991) Muscarinic acetylcholine receptor subtypes as agonist-dependent oncogenes. Proc Natl Acad Sci USA 88:4703-8

Hager J, Hansen L, Vaisse C, Vionnet N, Philippi A, Poller W, Velho G, Carcassi C, Contu L, Julier C (1995) A missense mutation in the glucagon receptor gene is associated with non-insulin-dependent diabetes mellitus. Nat Genet 9:299-304

Hall RA, Premont RT, Chow CW, Blitzer JT, Pitcher JA, Claing A, Stoffel RH, Barak LS, Shenolikar S, Weinman EJ, Grinstein S, Lefkowitz RJ (1998) The ß$_2$-adrenergic receptor interacts with the Na$^+$/H$^+$-exchanger regulatory factor to control Na$^+$/H$^+$ exchange. Nature 392:626-30

Hall RA, Premont RT, Lefkowitz RJ (1999) Heptahelical receptor signaling: beyond the G protein paradigm. J Cell Biol 145:927-32

Hamm HE (1998) The many faces of G protein signaling. J Biol Chem 273:669-72

Hani EH, Dupont S, Durand E, Dina C, Gallina S, Gantz I, Froguel P (2001) Naturally occurring mutations in the melanocortin receptor 3 gene are not associated with type 2 diabetes mellitus in French Caucasians. J Clin Endocrinol Metab 86:2895-8

Harteveld KL, Losekoot M, Fodde R, Giordano PC, Bernini LF (1997) The involvement of Alu repeats in recombination events at the α-globin gene cluster: characterization of two α°-thalassaemia deletion breakpoints. Hum Genet 99:528-34

Hatta N, Dixon C, Ray AJ, Phillips SR, Cunliffe WJ, Dale M, Todd C, Meggit S, Birch-MacHin MA, Rees JL (2001) Expression, candidate gene, and population studies of the melanocortin 5 receptor. J Invest Dermatol 116:564-70

Hausdorff WP, Caron MG, Lefkowitz RJ (1990) Turning off the signal: desensitization of ß-adrenergic receptor function. FASEB J 4:2881-9

Hayes JS, Lawler OA, Walsh MT, Kinsella BT (1999) The prostacyclin receptor is isoprenylated. Isoprenylation is required for efficient receptor-effector coupling. J Biol Chem 274:23707-18

Heard E, Clerc P, Avner P (1997) X-chromosome inactivation in mammals. Annu Rev Genet 31:571-610

Hebert TE, Moffett S, Morello JP, Loisel TP, Bichet DG, Barret C, Bouvier M (1996) A peptide derived from a ß_2-adrenergic receptor transmembrane domain inhibits both receptor dimerization and activation. J Biol Chem 271:16384-92

Heinflink M, Nussenzveig DR, Grimberg H, Lupu-Meiri M, Oron Y, Gershengorn MC (1995) A constitutively active mutant thyrotropin-releasing hormone receptor is chronically down-regulated in pituitary cells: evidence using chlordiazepoxide as a negative antagonist. Mol Endocrinol 9:1455-60

Heinonen P, Koulu M, Pesonen U, Karvonen MK, Rissanen A, Laakso M, Valve R, Uusitupa M, Scheinin M (1999) Identification of a three-amino acid deletion in the α_{2B}-adrenergic receptor that is associated with reduced basal metabolic rate in obese subjects. J Clin Endocrinol Metab 84:2429-33

Heisler LK, Chu HM, Brennan TJ, Danao JA, Bajwa P, Parsons LH, Tecott LH (1998) Elevated anxiety and antidepressant-like responses in serotonin 5-HT_{1A} receptor mutant mice. Proc Natl Acad Sci USA 95:15049-54

Helmreich EJ, Hofmann KP (1996) Structure and function of proteins in G-protein-coupled signal transfer. Biochim Biophys Acta 1286:285-322

Heuss C, Gerber U (2000) G-protein-independent signaling by G-protein-coupled receptors. Trends Neurosci 23:469-75

Heymann JAW, Subramaniam S (1997) Expression, stability, and membrane integration of truncation mutants of bovine rhodopsin. Proc Natl Acad Sci USA 94:4966-71

Hipkin RW, Liu X, Ascoli M (1995) Truncation of the C-terminal tail of the follitropin receptor does not impair the agonist- or phorbol ester-induced receptor phosphorylation and uncoupling. J Biol Chem 270:26683-9

Hirata T, Kakizuka A, Ushikubi F, Fuse I, Okuma M, Narumiya S (1994) Arg60 to Leu mutation of the human thromboxane A_2 receptor in a dominantly inherited bleeding disorder. J Clin Invest 94:1662-7

Ho HH, Du D, Gershengorn MC (1999) The N terminus of Kaposi's sarcoma-associated herpesvirus G protein-coupled receptor is necessary for high affinity chemokine binding but not for constitutive activity. J Biol Chem 274:31327-32

Hoffmann C, Moro S, Nicholas RA, Harden TK, Jacobson KA (1999) The role of amino acids in extracellular loops of the human $P2Y_1$ receptor in surface expression and activation processes. J Biol Chem 274:14639-47

Hollopeter G, Jantzen HM, Vincent D, Li G, England L, Ramakrishnan V, Yang RB, Nurden P, Nurden A, Julius D, Conley PB (2001) Identification of the platelet ADP receptor targeted by antithrombotic drugs. Nature 409:202-7

Hosoda K, Hammer RE, Richardson JA, Baynash AG, Cheung JC, Giaid A, Yanagisawa M (1994) Targeted and natural (piebald-lethal) mutations of endothelin-B receptor gene produce megacolon associated with spotted coat color in mice. Cell 79:1267-76

Hsu SY, Kudo M, Chen T, Nakabayashi K, Bhalla A, van der Spek PJ, van Duin M, Hsueh AJ (2000) The three subfamilies of leucine-rich repeat-containing G protein-coupled receptors (LGR): identification of LGR6 and LGR7 and the signaling mechanism for LGR7. Mol Endocrinol 14:1257-71

Hu LJ, Laporte J, Kress W, Kioschis P, Siebenhaar R, Poustka A, Fardeau M, Metzenberg A, Janssen EA, Thomas N, Mandel JL, Dahl N (1996) Deletions in Xq28 in two boys with myotubular myopathy and abnormal genital development define a new contiguous gene syndrome in a 430 kb region. Hum Mol Genet 5:139-43

Huang KS, Bayley H, Liao MJ, London E, Khorana HG (1981) Refolding of an integral membrane protein. Denaturation, renaturation, and reconstitution of intact bacteriorhodopsin and two proteolytic fragments. J Biol Chem 256:3802-9

Iida-Klein A, Guo J, Xie LY, Juppner H, Potts JT Jr, Kronenberg HM, Bringhurst FR, Abou-Samra AB, Segre GV (1995) Truncation of the carboxyl-terminal region of the rat parathyroid hormone (PTH)/PTH-related peptide receptor enhances PTH stimulation of adenylyl cyclase but not phospholipase C. J Biol Chem 270:8458-65

Iiri T, Bell SM, Baranski TJ, Fujita T, Bourne HR A (1999) $G_{s\alpha}$ mutant designed to inhibit receptor signaling through G_s. Proc Natl Acad Sci USA 96:499-504

Imamura F, Arimoto I, Fujiyoshi Y, Doi T (2000) W276 mutation in the endothelin receptor subtype B impairs G_q coupling but not G_i or G_o coupling. Biochemistry 39:686-92

Innamorati G, Sadeghi H, Birnbaumer M (1996) A fully active nonglycosylated V_2 vasopressin receptor. Mol Pharmacol 50:467-73

Inoue M, Hosoda K, Imura K, Kamata S, Fukuzawa M, Nakao K, Okada A (1998) Mutational analysis of the endothelin-B receptor gene in Japanese Hirschsprung's disease. J Pediatr Surg 33:1206-8

Ito M, Iwata N, Taniguchi T, Murayama T, Chihara K, Matsui T (1994) Functional characterization of two cholecystokinin-B/gastrin receptor isoforms: a preferential splice donor site in the human receptor gene. Cell Growth Differ 5:1127-35

Iwata N, Ozaki N, Inada T, Goldman D (2001) Association of a $5-HT_{5A}$ receptor polymorphism, Pro15Ser, to schizophrenia. Mol Psychiatry 6:217-9

Jacobson SG, Kemp CM, Cideciyan AV, Macke JP, Sung CH, Nathans J (1994) Phenotypes of stop codon and splice site rhodopsin mutations causing retinitis pigmentosa. Invest Ophthalmol Vis Sci 35:2521-34

Janssens R, Paindavoine P, Parmentier M, Boeynaems JM (1999) Human $P2Y_2$ receptor polymorphism: identification and pharmacological characterization of two allelic variants. Br J Pharmacol 127:709-16

Jaquette J, Segaloff DL (1997) Temperature sensitivity of some mutants of the lutropin/choriogonadotropin receptor. Endocrinology 138:85-91

Javitch JA, Fu D, Liapakis G, Chen J (1997) Constitutive activation of the ß2 adrenergic receptor alters the orientation of its sixth membrane-spanning segment. J Biol Chem 272:18546-9

Jin J, Mao GF, Ashby B (1997) Constitutive activity of human prostaglandin E receptor EP_3 isoforms. Br J Pharmacol 121:317-23

Jinnouchi H, Araki E, Miyamura N, Kishikawa H, Yoshimura R, Isami S, Yamaguchi K, Iwamatsu H, Shichiri M (1996) Analysis of vasopressin receptor type II (V_2R) gene in three Japanese pedigrees with congenital nephrogenic diabetes insipidus: identification of a family with complete deletion of the V_2R gene. Eur J Endocrinol 134:689-98

Jobert AS, Zhang P, Couvineau A, Bonaventure J, Roume J, Le Merrer M, Silve C (1998) Absence of functional receptors for parathyroid hormone and parathyroid hormone-related peptide in Blomstrand chondrodysplasia. J Clin Invest 102:34-40

Jones KA, Borowsky B, Tamm JA, Craig DA, Durkin MM, Dai M, Yao WJ, Johnson M, Gunwaldsen C, Huang LY, Tang C, Shen Q, Salon JA, Morse K, Laz T, Smith KE, Nagarathnam D, Noble SA, Branchek TA, Gerald C (1998) GABA$_B$ receptors function as a heteromeric assembly of the subunits GABA$_B$R1 and GABA$_B$R2. Nature 396:674-9

Jones PG, Curtis CAM, Hulme EC (1995) The function of a highly-conserved arginine residue in activation of the muscarinic M$_1$ receptor. Eur J Pharmacol 288:251-7

Jordan BA, Devi LA (1999) G-protein-coupled receptor heterodimerization modulates receptor function. Nature 399:697-700

Ju H, Venema VJ, Marrero MB, Venema RC (1998) Inhibitory interactions of the bradykinin B$_2$ receptor with endothelial nitric-oxide synthase. J Biol Chem 273:24025-9

Jung H, Windhaber R, Palm D, Schnackerz KD (1995) NMR and circular dichroism studies of synthetic peptides derived from the third intracellularloop of the ß-adrenoceptor. FEBS Lett 358:133-6

Kahn TW, Engelman DM (1992) Bacteriorhodopsin can be refolded from two independently stable transmembrane helices and the complementary five-helix fragment. Biochemistry 31:6144-51

Kaupmann K, Malitschek B, Schuler V, Heid J, Froestl W, Beck P, Mosbacher J, Bischoff S, Kulik A, Shigemoto R, Karschin A, Bettler B (1998) GABA$_B$-receptor subtypes assemble into functional heteromeric complexes. Nature 396:683-7

Kiel S, Bruss M, Bonisch H, Gothert M (2000) Pharmacological properties of the naturally occurring Phe-124-Cys variant of the human 5-HT$_{1B}$ receptor: changes in ligand binding, G-protein coupling and second messenger formation. Pharmacogenetics 10:655-66

Kilpatrick GJ, Dautzenberg FM, Martin GR, Eglen RM (1999) 7TM receptors: the splicing on the cake. Trends Pharmacol Sci 20:294-301

Klein C, Brin MF, Kramer P, Sena-Esteves M, de Leon D, Doheny D, Bressman S, Fahn S, Breakefield XO, Ozelius LJ (1999) Association of a missense change in the D2 dopamine receptor with myoclonus dystonia. Proc Natl Acad Sci USA 96:5173-6

Klein U, Ramirez MT, Kobilka BK, von Zastrow M (1997) A novel interaction between adrenergic receptors and the a-subunit of eukaryotic initiation factor 2B. J Biol Chem 272:19099-102

Kobilka BK, Kobilka TS, Daniel K, Regan JW, Caron MG, Lefkowitz RJ (1988) Chimeric α$_2$-,ß$_2$-adrenergic receptors: delineation of domains involved in effector coupling and ligand binding specificity. Science 240:1310-6

Koivisto UM, Palvimo JJ, Janne OA, Kontula K (1994) A single-base substitution in the proximal Sp1 site of the human low density lipoprotein receptor promoter as a cause of heterozygous familial hypercholesterolemia. Proc Natl Acad Sci USA 91:10526-30

Kolbe M, Besir H, Essen LO, Oesterhelt D (2000) Structure of the light-driven chloride pump halorhodopsin at 1.8 Å resolution. Science 288:1390-6

Kopin AS, McBride EW, Gordon MC, Quinn SM, Beinborn M (1997) Inter- and intraspecies polymorphisms in the cholecystokinin-B/gastrin receptor alter drug efficacy. Proc Natl Acad Sci USA 94:11043-8

Kopito RR (1999) Biosynthesis and degradation of CFTR. Physiol Rev 79:S167-73

Kopp P, van Sande J, Parma J, Duprez L, Gerber H, Joss E, Jameson JL, Dumont JE, Vassart G (1995) Brief report: congenital hyperthyroidism caused by a mutation in the thyrotropin-receptor gene. N Engl J Med 332:150-4

Kosugi S, Ban T, Akamizu T, Kohn LD (1992) Role of cysteine residues in the extracellular domain and exoplasmic loops of the transmembrane domain of the TSH receptor: effect of mutation to serine on TSH receptor activity and response to thyroid stimulating autoantibodies. Biochem Biophys Res Commun 189:1754-62

Kremer H, Kraaij R, Toledo SP, Post M, Fridman JB, Hayashida CY, van Reen M, Milgrom E, Ropers HH, Mariman E (1995) Male pseudohermaphroditism due to a homozygous missense mutation of the luteinizing hormone receptor gene. Nat Genet 9:160-4

Kroeger KM, Hanyaloglu AC, Seeber RM, Miles LE, Eidne KA (2001) Constitutive and agonist-dependent homo-oligomerization of the thyrotropin-releasing hormone receptor. Detection in living cells using bioluminescence resonance energy transfer. J Biol Chem 276:12736-43

Kunishima N, Shimada Y, Tsuji Y, Sato T, Yamamoto M, Kumasaka T, Nakanishi S, Jingami H, Morikawa K (2000) Structural basis of glutamate recognition by a dimeric metabotropic glutamate receptor. Nature 407:971-7

Laporte J, Biancalana V, Tanner SM, Kress W, Schneider V, Wallgren-Pettersson C, Herger F, Buj-Bello A, Blondeau F, Liechti-Gallati S, Mandel JL (2000) MTM1 mutations in X-linked myotubular myopathy. Hum Mutat 15:393-409

Lappalainen J, Zhang L, Dean M, Oz M, Ozaki N, Yu DH, Virkkunen M, Weight F, Linnoila M, Goldman D (1995) Identification, expression, and pharmacology of a Cys23-Ser23 substitution in the human 5-HT$_{2C}$ receptor gene (HTR2C). Genomics 27:274-9

Laue LL, Wu SM, Kudo M, Bourdony CJ, Cutler GB Jr, Hsueh AJ, Chan WY (1996) Compound heterozygous mutations of the luteinizing hormone receptor gene in Leydig cell hypoplasia. Mol Endocrinol 10:987-97

Laugwitz KL, Ungerer M, Schöneberg T, Weig HJ, Kronsbein K, Moretti A, Hoffmann K, Seyfarth M, Schultz G, Schömig A (1999) Adenoviral gene transfer of the human V$_2$ vasopressin receptor improves contractile force of rat cardiomyocytes. Circulation 99:925-33

Lawler OA, Miggin SM, Kinsella BT (2001) Protein kinase A mediated phosphorylation of serine 357 of the mouse prostacyclin receptor regulates its coupling to G$_s$-, to G$_i$- and to G$_q$-coupled effector signaling. J Biol Chem 276:33596-607

Layman LC, Cohen DP, Jin M, Xie J, Li Z, Reindollar RH, Bolbolan S, Bick DP, Sherins RR, Duck LW, Musgrove LC, Sellers JC, Neill JD (1998) Mutations in gonadotropin-releasing hormone receptor gene cause hypogonadotropic hypogonadism. Nat Genet 18:14-5

Le Gouill C, Parent JL, Rola-Pleszczynski M, Stankova J (1997) Role of the Cys[90], Cys[95] and Cys[173] residues in the structure and function of the human platelet-activating factor receptor. FEBS Lett 402:203-8

Le Gouill C, Parent JL, Caron CA, Gaudreau R, Volkov L, Rola-Pleszczynski M, Stankova J (1999) Selective modulation of wild type receptor functions by mutants of G-protein-coupled receptors. J Biol Chem 274:12548-54

Leavitt LM, Macaluso CR, Kim KS, Martin NP, Dumont ME (1999) Dominant negative mutations in the a-factor receptor, a G protein-coupled receptor encoded by the STE2 gene of the yeast Saccharomyces cerevisiae. Mol Gen Genet 261:917-32

Lee SP, O'Dowd BF, Ng GY, Varghese G, Akil H, Mansour A, Nguyen T, George SR (2000) Inhibition of cell surface expression by mutant receptors demonstrates that D_2 dopamine receptors exist as oligomers in the cell. Mol Pharmacol 58:120-8

Lee TW, Cotecchia S, Milligan G (1997) Up-regulation of the levels of expression and function of a constitutively active mutant of the hamster α_{1B}-adrenoceptor by ligands that act as inverse agonists. Biochem J 325:733-9

Lefkowitz RJ, Cotecchia S, Samama P, Costa T (1993) Constitutive activity of receptors coupled to guanine nucleotide regulatory proteins. Trends Pharmacol Sci 14:303-7

Li S, Liu X, Ascoli M (2000) p38JAB1 binds to the intracellular precursor of the lutropin/choriogonadotropin receptor and promotes its degradation. J Biol Chem 275:13386-93

Lin L, Faraco J, Li R, Kadotani H, Rogers W, Lin X, Qiu X, de Jong PJ, Nishino S, Mignot E (1999) The sleep disorder canine narcolepsy is caused by a mutation in the hypocretin (orexin) receptor 2 gene. Cell 98:365-76

Lin SC, Lin CR, Gukovsky I, Lusis AJ, Sawchenko PE, Rosenfeld MG (1993) Molecular basis of the little mouse phenotype and implications for cell type-specific growth. Nature 364:208-13

Lin Z, Shenker A, Pearlstein R (1997) A model of the lutropin/choriogonadotropin receptor: insights into the structural and functional effects of constitutively activating mutations. Protein Eng 10:501-10

Liu F, Wan Q, Pristupa ZB, Yu XM, Wang YT, Niznik HB (2000) Direct protein-protein coupling enables cross-talk between dopamine D5 and gamma-aminobutyric acid A receptors. Nature 403:274-80

Liu T, DeCostanzo AJ, Liu X, Wang Hy, Hallagan S, Moon RT, Malbon CC (2001) G protein signaling from activated rat frizzled-1 to the beta-catenin-Lef-Tcf pathway. Science 292:1718-22

Liu G, Duranteau L, Carel JC, Monroe J, Doyle DA, Shenker A (1999) Leydig-cell tumors caused by an activating mutation of the gene encoding the luteinizing hormone receptor. N Engl J Med 341:1731-6

Liu IS, Seeman P, Sanyal S, Ulpian C, Rodgers-Johnson PE, Serjeant GR, Van Tol HH (1996) Dopamine D4 receptor variant in Africans, D4valine194glycine, is insensitive to dopamine and clozapine: report of a homozygous individual. Am J Med Genet 61:277-82

Liu IS, Kusumi I, Ulpian C, Tallerico T, Seeman P (1998) A serotonin-4 receptor-like pseudogene in humans. Brain Res Mol Brain Res 53:98-103

Liu J, Schöneberg T, van Rhee M, Wess J (1995) Mutational analysis of the relative orientation of transmembrane helices I and VII in G protein-coupled receptors. J Biol Chem 270:19532-9

Liu J, Wess J (1996) Different single receptor domains determine the distinct G protein coupling profiles of members of the vasopressin receptor family. J Biol Chem 271:8772-8

Liu R, Paxton WA, Choe S, Ceradini D, Martin SR, Horuk R, MacDonald ME, Stuhlmann H, Koup RA, Landau NR (1996) Homozygous defect in HIV-1 coreceptor accounts for resistance of some multiply-exposed individuals to HIV-1 infection. Cell 86:367-77

Liu T, DeCostanzo AJ, Liu X, Wang Hy, Hallagan S, Moon RT, Malbon CC (2001) G protein signaling from activated rat frizzled-1 to the ß-catenin-Lef-Tcf pathway. Science 292:1718-22

Lohmann DR (1999) RB1 gene mutations in retinoblastoma. Hum Mutat 14:283-8

Lu D, Vage DI, Cone RD (1998) A ligand-mimetic model for constitutive activation of the melanocortin-1 receptor. Mol Endocrinol 12:592-604

Luecke H, Schobert B, Richter HT, Cartailler JP, Lanyi JK (1999) Structure of bacteriorhodopsin at 1.55 Å resolution. J Mol Biol 291:899-911

Luzius H, Jans DA, Grunbaum EG, Moritz A, Rascher W, Fahrenholz F (1992) A low affinity vasopressin V2-receptor in inherited nephrogenic diabetes insipidus. J Recept Res 12:351-68

Macke JP, Davenport CM, Jacobson SG, Hennessey JC, Gonzalez-Fernandez F, Conway BP, Heckenlively J, Palmer R, Maumenee IH, Sieving P (1993) Identification of novel rhodopsin mutations responsible for retinitis pigmentosa: implications for the structure and function of rhodopsin. Am J Hum Genet 53:80-9

Maggio R, Vogel Z, Wess J (1993) Co-expression studies with mutant muscarinic/adrenergic receptors provide evidence for intermolecular crosstalk between G protein-linked receptors. Proc Natl Acad Sci USA 90:3103-7

Manivet P, Mouillet-Richard S, Callebert J, Nebigil CG, Maroteaux L, Hosoda S, Kellermann O, Launay JM (2000) PDZ-dependent activation of nitric-oxide synthases by the serotonin 2B receptor. J Biol Chem 275:9324-31

Marchese A, Nguyen T, Malik P, Xu S, Cheng R, Xie Z, Heng HH, George SR, Kolakowski LF Jr, O'Dowd BF (1998) Cloning genes encoding receptors related to chemoattractant receptors. Genomics 50:281-6

Margeta-Mitrovic M, Jan YN, Jan LY (2000) A trafficking checkpoint controls GABA_B receptor heterodimerization. Neuron 27:97-106

Marie J, Koch C, Pruneau D, Paquet JL, Groblewski T, Larguier R, Lombard C, Deslauriers B, Maigret B, Bonnafous JC (1999) Constitutive activation of the human bradykinin B_2 receptor induced by mutations in transmembrane helices III and VI. Mol Pharmacol 55:92-101

Martin MM, Wu SM, Martin AL, Rennert OM, Chan WY (1998) Testicular seminoma in a patient with a constitutively activating mutation of the luteinizing hormone/chorionic gonadotropin receptor. Eur J Endocrinol 139:101-6

Mason DA, Moore JD, Green SA, Liggett SB (1999) A gain-of-function polymorphism in a G-protein coupling domain of the human ß_1-adrenergic receptor. J Biol Chem 274:12670-4

McLatchie LM, Fraser NJ, Main MJ, Wise A, Brown J, Thompson N, Solari R, Lee MG, Foord SM (1998) RAMPs regulate the transport and ligand specificity of the calcitonin-receptor-like receptor. Nature 393:333-9

Milligan G, Bond RA (1997) Inverse agonism and the regulation of receptor number. Trends Pharmacol Sci 18:468-74

Milligan G, Bond RA, Lee M (1995) Inverse agonism: pharmacological curiosity or potential therapeutic strategy? Trends Pharmacol Sci 16:10-3

Misrahi M, Meduri G, Pissard S, Bouvattier C, Beau I, Loosfelt H, Jolivet A, Rappaport R, Milgrom E, Bougneres P (1997) Comparison of immunocytochemical and molecular features with the phenotype in a case of incomplete male pseudohermaphroditism associated with a mutation of the luteinizing hormone receptor. J Clin Endocrinol Metab 82:2159-65

Morello JP, Salahpour A, Laperriere A, Bernier V, Arthus MF, Lonergan M, Petaja-Repo U, Angers S, Morin D, Bichet DG, Bouvier M (2000) Pharmacological chaperones rescue cell-surface expression and function of misfolded V_2 vasopressin receptor mutants. J Clin Invest 105:887-95

Morello JP, Salahpour A, Petaja-Repo UE, Laperriere A, Lonergan M, Arthus MF, Nabi IR, Bichet DG, Bouvier M (2001) Association of calnexin with wild type and mutant AVPR2 that cause nephrogenic diabetes insipidus. Biochemistry 40:6766-75

Morimura H, Saindelle-Ribeaudeau F, Berson EL, Dryja TP 1999 Mutations in RGR, encoding a light-sensitive opsin homologue, in patients with retinitis pigmentosa. Nat Genet 23:393-4

Moses AM, Sangani G, Miller JL (1995) Proposed cause of marked vasopressin resistance in a female with an X-linked recessive V2 receptor abnormality. J Clin Endocrinol Metab 80:1184-6

Mullikin JC, Hunt SE, Cole CG, Mortimore BJ, Rice CM, Burton J, Matthews LH, Pavitt R, Plumb RW, Sims SK, Ainscough RM, Attwood J, Bailey JM, Barlow K, Bruskiewich RM, Butcher PN, Carter NP, Chen Y, Clee CM, Coggill PC, Davies J, Davies RM, Dawson E, Francis MD, Joy AA, Lamble RG, Langford CF, Macarthy J, Mall V, Moreland A, Overton-Larty EK, Ross MT, Smith LC, Steward CA, Sulston JE, Tinsley EJ, Turney KJ, Willey DL, Wilson GD, McMurray AA, Dunham I, Rogers J, Bentley DR (2000) An SNP map of human chromosome 22. Nature 407:516-20

Nagase T, Seki N, Tanaka A, Ishikawa K, Nomura N (1995) Prediction of the coding sequences of unidentified human genes. IV. The coding sequences of 40 new genes (KIAA0121-KIAA0160) deduced by analysis of cDNA clones from human cell line KG-1. DNA Res 2:167-74

Nakagawa M, Miyamoto T, Kusakabe R, Takasaki S, Takao T, Shichida Y, Tsuda M (2001) O-Glycosylation of G-protein-coupled receptor, octopus rhodopsin. Direct analysis by FAB mass spectrometry. FEBS Lett 496:19-24

Namba T, Sugimoto Y, Negishi M, Irie A, Ushikubi F, Kakizuka A, Ito S, Ichikawa A, Narumiya S (1993) Alternative splicing of C-terminal tail of prostaglandin E receptor subtype EP$_3$ determines G-protein specificity. Nature 365:166-70

Nathans J, Hogness DS (1983) Isolation, sequence analysis, and intron-exon arrangement of the gene encoding bovine rhodopsin. Cell 34:807-14

Naville D, Barjhoux L, Jaillard C, Faury D, Despert F, Esteva B, Durand P, Saez JM, Begeot M (1996) Demonstration by transfection studies that mutations in the adrenocorticotropin receptor gene are one cause of the hereditary syndrome of glucocorticoid deficiency. J Clin Endocrinol Metab 81:1442-8

Nelson G, Hoon MA, Chandrashekar J, Zhang Y, Ryba NJP, Zuker CS (2001) Mammalian sweet taste receptors. Cell 106: 381-90

Nielsen SM, Elling CE, Schwartz TW (1998) Split-receptors in the tachykinin neurokinin-1 system--mutational analysis of intracellular loop 3. Eur J Biochem 251:217-26

Nijenhuis WA, Oosterom J, Adan RA (2001) AgRP(83-132) acts as an inverse agonist on the human-melanocortin-4 receptor. Mol Endocrinol 15:164-71

Nimchinsky EA, Hof PR, Janssen WGM, Morrison JH, Schmauss C (1997) Expression of dopamine D$_3$ receptor dimers and tetramers in brain and in transfected cells. J Biol Chem 272:29229-37

Niswender CM, Copeland SC, Herrick-Davis K, Emeson RB, Sanders-Bush E (1999) RNA editing of the human serotonin 5-hydroxytryptamine 2C receptor silences constitutive activity. J Biol Chem 274:9472-8

Nomura Y, Onigata K, Nagashima T, Yutani S, Mochizuki H, Nagashima K, Morikawa A (1997) Detection of skewed X-inactivation in two female carriers of vasopressin type 2 receptor gene mutation. J Clin Endocrinol Metab 82:3434-7

Nothen MM, Cichon S, Hemmer S, Hebebrand J, Remschmidt H, Lehmkuhl G, Poustka F, Schmidt M, Catalano M, Fimmers R (1994) Human dopamine D$_4$ receptor gene: frequent occurrence of a null allele and observation of homozygosity. Hum Mol Genet 3:2207-12

Offermanns S (2000) Mammalian G-protein function in vivo: new insights through altered gene expression. Rev Physiol Biochem Pharmacol 140:63-133

Okuda-Ashitaka E, Sakamoto K, Ezashi T, Miwa K, Ito S, Hayaishi O (1996) Suppression of prostaglandin E receptor signaling by the variant form of EP$_1$ subtype. J Biol Chem 271:31255-61

Oldenhof J, Vickery R, Anafi M, Oak J, Ray A, Schoots O, Pawson T, von Zastrow M, Van Tol HH (1998) SH3 binding domains in the dopamine D$_4$ receptor. Biochemistry 37:15726-36

Okada T, Ernst OP, Palczewski K, Hofmann KP (2001) Activation of rhodopsin: new insights from structural and biochemical studies. Trends Biochem Sci 26:318-24

Oliveira SA, Shenk TE (2001) Murine cytomegalovirus M78 protein, a G protein-coupled receptor homologue, is a constituent of the virion and facilitates accumulation of immediate-early viral mRNA. Proc Natl Acad Sci USA 98:3237-42

Otaki JM, Firestein S (2001) Length analyses of mammalian g-protein-coupled receptors. J Theor Biol 211:77-100

Overton MC, Blumer KJ (2000) G-protein-coupled receptors function as oligomers in vivo. Curr Biol 10:341-4

Pagano A, Rovelli G, Mosbacher J et al. (2001) C-terminal interaction is essential for surface trafficking but not for heteromeric assembly of GABA$_B$ receptors. J Neuroscience 21:1189-202

Palczewski K, Kumasaka T, Hori T, Behnke CA, Motoshima H, Fox BA, Le Trong I, Teller DC, Okada T, Stenkamp RE, Yamamoto M, Miyano M (2000) Crystal structure of rhodopsin: A G protein-coupled receptor. Science 289:739-45

Park YS, Lee YS, Cho NJ, Kaang BK (2000) Alternative splicing of gar-1, a Caenorhabditis elegans G-protein-linked acetylcholine receptor gene. Biochem Biophys Res Commun 268:354-8

Parma J, Duprez L, van Sande J, Cochaux P, Gervy C, Mockel J, Dumont J, Vassart G (1993) Somatic mutations in the thyrotropin receptor gene cause hyperfunctioning thyroid adenomas. Nature 365:649-51

Parma J, Duprez L, Van Sande J, Hermans J, Rocmans P, Van Vliet G, Costagliola S, Rodien P, Dumont JE, Vassart G (1997) Diversity and prevalence of somatic mutations in the thyrotropin receptor and G$_s\alpha$ genes as a cause of toxic thyroid adenomas. J Clin Endocrinol Metab 82:2695-701

Pasel K, Schulz A, Timmermann K, Linnemann K, Hoeltzenbein M, Jääskeläinen J, Grüters A, Filler G, Schöneberg T (2000) Functional characterization of the molecular defects causing nephrogenic diabetes insipidus in eight families. J Clin Endocrinol Metab 85:1703-10

Perez DM, Hwa J, Gaivin R, Mathur M, Brown F, Graham RM (1996) Constitutive activation of a single effector pathway: evidence for multiple activation states of a G protein-coupled receptor. Mol Pharmacol 49:112-22

Perlman JH, Wang W, Nussenzveig DR, Gershengorn MC (1995) A disulfide bond between conserved extracellular cysteines in the thyrotropin-releasing hormone receptor is critical for binding. J Biol Chem 270:24682-5

Perlman JH, Colson AO, Wang W, Bence K, Osman R, Gershengorn MC (1997) Interactions between conserved residues in transmembrane helices 1, 2, and 7 of the thyrotropin-releasing hormone receptor. J Biol Chem 272:11937-42

Peroutka SJ, Howell TA (1994) The molecular evolution of G protein-coupled receptors: focus on 5-hydroxytryptamine receptors. Neuropharmacology 33:319-24

Pierce KL, Regan JW (1998) Prostanoid receptor heterogeneity through alternative mRNA splicing. Life Sci 62:1479-83

Pitteloud N, Boepple PA, DeCruz S, Valkenburgh SB, Crowley WF Jr, Hayes FJ (2001) The fertile eunuch variant of idiopathic hypogonadotropic hypogonadism: spontaneous reversal associated with a homozygous mutation in the gonadotropin-releasing hormone receptor. J Clin Endocrinol Metab 86:2470-5

Podesta EJ, Solano AR, Attar R, Sanchez ML, Molina Y, Vedia L (1983) Receptor aggregation induced by antilutropin receptor antibody and biological response in rat testis Leydig cells. Proc Natl Acad Sci USA 80:3986-90

Pollak MR, Brown EM, Chou YH, Hebert SC, Marx SJ, Steinmann B, Levi T, Seidman CE, Seidman JG (1993) Mutations in the human Ca^{2+}-sensing receptor gene cause familial hypocalciuric hypercalcemia and neonatal severe hyperparathyroidism. Cell 75:1297-303

Porter JE, Hwa J, Perez DM (1996) Activation of the α_{1b}-adrenergic receptor is initiated by disruption of an interhelical salt bridge constraint. J Biol Chem 271:28318-23

Prezeau L, Gomeza J, Ahern S, Mary S, Galvez T, Bockaert J, Pin JP (1996) Changes in the carboxyl-terminal domain of metabotropic glutamate receptor 1 by alternative splicing generate receptors with differing agonist-independent activity. Mol Pharmacol 49:422-9

Puffenberger EG, Hosoda K, Washington SS, Nakao K, deWit D, Yanagisawa M, Chakravart A (1994) A missense mutation of the endothelin-B receptor gene in multigenic Hirschsprung's disease. Cell 79:1257-66

Rana BK, Shiina T, Insel PA (2001) Genetic variations and polymorphisms of g protein-coupled receptors: functional and therapeutic implications. Ann Rev Pharmacol Toxicol 41: 593-624.

Rando TA, Disatnik MH, Zhou LZ (2000) Rescue of dystrophin expression in mdx mouse muscle by RNA/DNA oligonucleotides. Proc Natl Acad Sci USA 97:5363-8

Rands E, Candelore MR, Cheung AH, Hill WS, Strader CD, Dixon RA (1990) Mutational analysis of ß-adrenergic receptor glycosylation. J Biol Chem 265:10759-64

Rao VR, Cohen GB, Oprian DD (1994) Rhodopsin mutation G90D and a molecular mechanism for congenital night blindness. Nature 367:639-42

Rasmussen SG, Jensen AD, Liapakis G, Ghanouni P, Javitch JA, Gether U (1999) Mutation of a highly conserved aspartic acid in the β_2 adrenergic receptor: constitutive activation, structural instability, and conformational rearrangement of transmembrane segment 6. Mol Pharmacol 56:175-84

Ray K, Clapp P, Goldsmith PK, Spiegel AM (1998) Identification of the sites of N-linked glycosylation on the human calcium receptor and assessment of their role in cell surface expression and signal transduction. J Biol Chem 273:34558-67

Redfern CH, Coward P, Degtyarev MY, Lee EK, Kwa AT, Hennighausen L, Bujard H, Fishman GI, Conklin BR (1999) Conditional expression and signaling of a specifically designed G_i-coupled receptor in transgenic mice. Nat Biotechnol 17:165-9

Reig C, Alvarez AI, Tejada I, Molina M, Arostegui E, Martin R, Antich J, Carballo M (1996) New mutation in the 3'-acceptor splice site of intron 4 in the rhodopsin gene associated with autosomal dominant retinitis pigmentosa in a Basque family. Hum Mutat 8:93-4

Ren Q, Kurose H, Lefkowitz RJ, Cotecchia S (1993) Constitutively active mutants of the α_2-adrenergic receptor. J Biol Chem 268:16483-7

Restagno G, Maghtheh M, Bhattacharya S, Ferrone M, Garnerone S, Samuelly R, Carbonara A (1993) A large deletion at the 3' end of the rhodopsin gene in an Italian family with a diffuse form of autosomal dominant retinitis pigmentosa. Hum Mol Genet 2:207-8

Rhee MH, Nevo I, Levy R, Vogel Z (2000) Role of the highly conserved Asp-Arg-Tyr motif in signal transduction of the CB₂ cannabinoid receptor. FEBS Lett 466:300-4

Ridge KD, Lee SS, Yao LL (1995) In vivo assembly of rhodopsin from expressed polypeptide fragments. Proc Natl Acad Sci USA 92:3204-8

Riek RP, Rigoutsos I, Novotny J, Graham RM (2001) Non-α-helical elements modulate polytopic membrane protein architecture. J Mol Biol 306:349-62

Robinson PR, Cohen GB, Zhukovsky EA, Oprian DD (1992) Constitutively active mutants of rhodopsin. Neuron 9:719-25

Rocheville M, Lange DC, Kumar U, Patel SC, Patel RC, Patel YC (2000a) Receptors for dopamine and somatostatin: formation of hetero-oligomers with enhanced functional activity. Science 288:154-7

Rocheville M, Lange DC, Kumar U, Sasi R, Patel RC, Patel YC (2000b) Subtypes of the somatostatin receptor assemble as functional homo- and heterodimers. J Biol Chem 275:7862-9

Romano C, Yang WL, O'Malley KL (1996) Metabotropic glutamate receptor 5 is a disulfide-linked dimer. J Biol Chem 271:28612-6

Rose PM, Lynch JS, Frazier ST, Fisher SM, Chung W, Battaglino P, Fathi Z, Leibel R, Fernandes P (1997) Molecular genetic analysis of a human neuropeptide Y receptor. The human homolog of the murine "Y5" receptor may be a pseudogene. J Biol Chem 272:3622-7

Rosenkilde MM, Kledal TN, Brauner-Osborne H, Schwartz TW (1999) Agonists and inverse agonists for the herpesvirus 8-encoded constitutively active seven-transmembrane oncogene product, ORF-74. J Biol Chem 274:956-61

Rosenthal W, Seibold A, Antaramian A, Lonergan M, Arthus MF, Hendy GN, Birnbaumer M, Bichet DG (1992) Molecular identification of the gene responsible for congenital nephrogenic diabetes insipidus. Nature 359:233-5

Rotondo A, Nielsen DA, Nakhai B, Hulihan-Giblin B, Bolos A, Goldman D (1997) Agonist-promoted down-regulation and functional desensitization in two naturally occurring variants of the human serotonin1A receptor. Neuropsychopharmacology 17:18-26

Royant A, Nollert P, Edman K, Neutze R, Landau EM, Pebay-Peyroula E, Navarro J (2001) X-ray structure of sensory rhodopsin II at 2.1-Å resolution. Proc Natl Acad Sci USA 98:10131-6

Rozell TG, Davis DP, Chai Y, Segaloff DL (1998) Association of gonadotropin receptor precursors with the protein folding chaperone calnexin. Endocrinology 139:1588-93

Rubinstein M, Phillips TJ, Bunzow JR, Falzone TL, Dziewczapolski G, Zhang G, Fang Y, Larson JL, McDougall JA, Chester JA, Saez C, Pugsley TA, Gershanik O, Low MJ, Grandy DK (1997) Mice lacking dopamine D₄ receptors are supersensitive to ethanol, cocaine, and methamphetamine. Cell 90:991-1001

Russo D, Arturi F, Schlumberger M, Caillou B, Monier R, Filetti S, Suarez HG (1995) Activating mutations of the TSH receptor in differentiated thyroid carcinomas. Oncogene 11:1907-11

Russo D, Arturi F, Suarez HG, Schlumberger M, Du Villard JA, Crocetti U, Filetti S (1996) Thyrotropin receptor gene alterations in thyroid hyperfunctioning adenomas. J Clin Endocrinol Metab 81:1548-51

Sadeghi H, Birnbaumer M (1999) O-Glycosylation of the V_2 vasopressin receptor. Glycobiology 9:731-7

Sadeghi HM, Innamorati G, Birnbaumer M (1997a) An X-linked NDI mutation reveals a requirement for cell surface V_2R expression. Mol Endocrinol 11:706-13

Sadeghi H, Robertson GL, Bichet DG, Innamorati G, Birnbaumer M (1997b) Biochemical basis of partial nephrogenic diabetes insipidus phenotypes. Mol Endocrinol 11:1806-13

Samama P, Bond RA, Rockman HA, Milano CA, Lefkowitz RJ (1997) Ligand-induced overexpression of a constitutively active ß$_2$-adrenergic receptor: pharmacological creation of a phenotype in transgenic mice. Proc Natl Acad Sci USA 94:137-41

Samson M, Libert F, Doranz BJ, Rucker J, Liesnard C, Farber CM, Saragosti S, Lapoumeroulie C, Cognaux J, Forceille C, Muyldermans G, Verhofstede C, Burtonboy G, Georges M, Imai T, Rana S, Yi Y, Smyth RJ, Collman RG, Doms RW, Vassart G, Parmentier M (1996) Resistance to HIV-1 infection in caucasian individuals bearing mutant alleles of the CCR-5 chemokine receptor gene. Nature 382:722-5

Sato S, Ward C, Krouse ME, Wine JJ, Kopito RR (1996) Glycerol reverses the misfolding phenotype of the most common cystic fibrosis mutation. J Biol Chem 271:635-8

Savarese TM, Wang CD, Fraser CM (1992) Site-directed mutagenesis of the rat m1 muscarinic acetylcholine receptor. Role of conserved cysteines in receptor function. J Biol Chem 267:11439-48

Scheer A, Fanelli F, Costa T, De Benedetti PG, Cotecchia S (1996) Constitutively active mutants of the α_{1B}-adrenergic receptor: role of highly conserved polar amino acids in receptor activation. EMBO J 15:3566-78

Scheer A, Fanelli F, Costa T, De Benedetti PG, Cotecchia S (1997) The activation process of the α_{1B}-adrenergic receptor: potential role of protonation and hydrophobicity of a highly conserved aspartate. Proc Natl Acad Sci USA 94:808-13

Schipani E, Kruse K, Juppner H (1995) A constitutively active mutant PTH-PTHrP receptor in Jansen-type metaphyseal chondrodysplasia. Science 268:98-100

Schipani E, Lanske B, Hunzelman J, Luz A, Kovacs CS, Lee K, Pirro A, Kronenberg HM, Juppner H (1997) Targeted expression of constitutively active receptors for parathyroid hormone and parathyroid hormone-related peptide delays endochondral bone formation and rescues mice that lack parathyroid hormone-related peptide. Proc Natl Acad Sci USA 94:13689-94

Schöneberg T, Liu J, Wess J (1995) Plasma membrane localization and functional rescue of truncated forms of a G protein-coupled receptor. J Biol Chem 270:18000-6

Schöneberg T, Yun J, Wenkert D, Wess J (1996) Functional rescue of mutant V_2 vasopressin receptors causing nephrogenic diabetes insipidus by a co-expressed receptor polypeptide. EMBO J 15:1283-91

Schöneberg T, Sandig V, Wess J, Gudermann T, Schultz G (1997) Reconstitution of mutant V_2 vasopressin receptors by adenovirus-mediated gene transfer. Molecular basis and clinical implication. J Clin Invest 100:1547-56

Schöneberg T, Schulz A, Biebermann H, Grüters A, Grimm T, Hübschmann K, Filler G, Gudermann T, Schultz G (1998) V_2 vasopressin receptor dysfunction in nephrogenic diabetes insipidus caused by different molecular mechanisms. Hum Mutat 12:196-205

Schöneberg T, Pasel K, von Baehr V, Schulz A, Volk HD, Gudermann T, Filler G (1999) Compound deletion of the rhoGAP C1 and V_2 vasopressin receptor genes in a patient with nephrogenic diabetes insipidus. Hum Mutat 14:163-74

Schülein R, Hermosilla R, Oksche A, Dehe M, Wiesner B, Krause G, Rosenthal W (1998) A dileucine sequence and an upstream glutamate residue in the intracellular carboxyl terminus of the vasopressin V_2 receptor are essential for cell surface transport in COS.M6 cells. Mol Pharmacol 54:525-35

Schülein R, Zühlke K, Krause G, Rosenthal W (2000) Functional rescue of three vasopressin V_2 receptor mutants causing nephrogenic diabetes insipidus by a second site suppressor mutation. J Biol Chem 276:8384-92

Schulz A, Schöneberg T, Paschke R, Schultz G, Gudermann T (1999) Role of the third intracellular loop for the activation of gonadotropin receptors. Mol Endocrinol 13:181-90

Schulz A, Bruns C, Henklein P, Krause G, Schubert M, Gudermann T, Wray V, Schultz G, Schöneberg T (2000a) Requirement of specific intrahelical interactions for stabilizing the inactive conformation of glycoprotein hormone receptors. J Biol Chem 275: 37860-9

Schulz A, Grosse R, Schultz G, Gudermann T, Schöneberg T (2000b) Structural implication for receptor oligomerization from functional reconstitution studies of mutant V_2 vasopressin receptors. J Biol Chem 275:2381-9

Schulz A, Schöneberg T (2000c) Analysis of molecular mechanisms causing V_2 vasopressin receptor dysfunction in 8 NDI families. NDI Conference – La Jolla, March 10-12

Schwarzler A, Kreienkamp HJ, Richter D (2000) Interaction of the somatostatin receptor subtype 1 with the human homolog of the Shk1 kinase-binding protein from yeast. J Biol Chem 275:9557-62

Seeman P, Nam D, Ulpian C, Liu IS, Tallerico T (2000) New dopamine receptor, $D_{2(Longer)}$, with unique TG splice site, in human brain. Brain Res Mol Brain Res 76:132-41

Seibold A, Dagarag M, Birnbaumer M (1998) Mutations of the DRY motif that preserve ß$_2$-adrenoceptor coupling. Receptors Channels 5:375-85

Sheikh SP, Zvyaga TA, Lichtarge O, Sakmar TP, Bourne HR (1996) Rhodopsin activation blocked by metal-ion-binding sites linking transmembrane helices C and F. Nature 383:347-50

Shenker A, Laue L, Kosugi S, Merendino JJ Jr, Minegishi T, Cutler GB Jr (1993) A constitutively activating mutation of the luteinizing hormone receptor in familial male precocious puberty. Nature 365:652-4

Shibata K, Hirasawa A, Moriyama N, Kawabe K, Ogawa S, Tsujimoto G (1996) a_{1a}-adrenoceptor polymorphism: pharmacological characterization and association with benign prostatic hypertrophy. Br J Pharmacol 118:1403-8

Shima Y, Tsukada T, Nakanishi K, Ohta H (1998) Association of the Trp64Arg mutation of the ß$_3$-adrenergic receptor with fatty liver and mild glucose intolerance in Japanese subjects. Clin Chim Acta 274:167-76

Sims KB, Ozelius L, Corey T, Rinehart WB, Liberfarb R, Haines J, Chen WJ, Norio R, Sankila E, de la Chapelle A (1989) Norrie disease gene is distinct from the monoamine oxidase genes. Am J Hum Genet 45:424-34

Sina M, Hinney A, Ziegler A, Neupert T, Mayer H, Siegfried W, Blum WF, Remschmidt H, Hebebrand J (1999) Phenotypes in three pedigrees with autosomal dominant obesity caused by haploinsufficiency mutations in the melano-cortin-4 receptor gene. Am J Hum Genet 65:1501-7

Singh N, Zanusso G, Chen SG, Fujioka H, Richardson S, Gambetti P, Petersen RB (1997) Prion protein aggregation reverted by low temperature in transfected cells carrying a prion protein gene mutation. J Biol Chem 272:28461-70

Small K, Iber J, Warren ST (1997) Emerin deletion reveals a common X-chromosome inversion mediated by inverted repeats. Nat Genet 16:96-9

Small KM, Forbes SL, Bridges K, Liggett SB (2000) An Asn to Lys polymorphism in the third intracellular loop of the human α_{2A}-adrenergic receptor imparts enhanced agonist-promoted G_i coupling. J Biol Chem 275:38518-23

Small KM, Brown KM, Forbes SL, Liggett SB (2001) Modification of the ß$_2$-adrenergic receptor to engineer a receptor-effector complex for gene therapy. J Biol Chem 276:31596-601

Smit MJ, Leurs R, Alewijnse AE, Blauw J, Van Nieuw Amerongen GP, Van De Vrede Y, Roovers E, Timmerman H (1996) Inverse agonism of histamine H_2 antagonist accounts for upregulation of spontaneously active histamine H_2 receptors. Proc Natl Acad Sci USA 93:6802-7

Smith MW, Dean M, Carrington M, Winkler C, Huttley GA, Lomb DA, Goedert JJ, O'Brien TR, Jacobson LP, Kaslow R, Buchbinder S, Vittinghoff E, Vlahov D, Hoots K, Hilgartner MW, O'Brien SJ (1997) Contrasting genetic influence of CCR2 and CCR5 variants on HIV-1 infection and disease progression. Hemophilia Growth and Development Study (HGDS), Multicenter AIDS Cohort Study (MACS), Multicenter Hemophilia Cohort Study (MHCS), San Francisco City Cohort (SFCC), ALIVE Study. Science 277:959-65

Sobell JL, Lind TJ, Sigurdson DC, Zald DH, Snitz BE, Grove WM, Heston LL, Sommer SS (1995) The D_5 dopamine receptor gene in schizophrenia: identification of a nonsense change and multiple missense changes but lack of association with disease. Hum Mol Genet 4:507-14

Stacey M, Lin HH, Gordon S, McKnight AJ (2000) LNB-TM7, a group of seven-transmembrane proteins related to family-B G-protein-coupled receptors. Trends Biochem Sci 25:284-9

Stein SA, Oates EL, Hall CR, Grumbles RM, Fernandez LM, Taylor NA, Puett D, Jin S (1994) Identification of a point mutation in the thyrotropin receptor of the hyt/hyt hypothyroid mouse. Mol Endocrinol 8:129-38

Stühmer W, Conti F, Suzuki H, Wang XD, Noda M, Yahagi N, Kubo H, Numa S (1989) Structural parts involved in activation and inactivation of the sodium channel. Nature 339:597-603

Sullivan LJ, Makris GS, Dickinson P, Mulhall LE, Forrest S, Cotton RG, Loughnan MS (1993) A new codon 15 rhodopsin gene mutation in autosomal dominant retinitis pigmentosa is associated with sectorial disease. Arch Ophthalmol 111:1512-7

Sung CH, Schneider BG, Agarwal N, Papermaster DS, Nathans J (1991) Functional heterogeneity of mutant rhodopsins responsible for autosomal dominant retinitis pigmentosa. Proc Natl Acad Sci USA 88:8840-4

Sunthornthepvarakui T, Gottschalk ME, Hayashi Y, Refetoff S (1995) Brief report: resistance to thyrotropin caused by mutations in the thyrotropin-receptor gene. N Engl J Med 332:155-60

Surprenant A, Horstman DA, Akbarali H, Limbird LE (1992) A point mutation of the ß$_2$-adrenoceptor that blocks coupling to potassium but not calcium currents. Science 257:977-80

Taboulet J, Frenkian M, Frendo JL, Feingold N, Jullienne A, de Vernejoul MC (1998) Calcitonin receptor polymorphism is associated with a decreased fracture risk in post-menopausal women. Hum Mol Genet 7:2129-33

Tajima T, Nakae J, Takekoshi Y, Takahashi Y, Yuri K, Nagashima T, Fujieda K (1996) Three novel AVPR2 mutations in three Japanese families with X-linked nephrogenic diabetes insipidus. Pediatr Res 39:522-6

Takagi Y, Ninomiya H, Sakamoto A, Miwa S, Masaki T (1995) Structural basis of G protein specificity of human endothelin receptors. A study with endothelinA/B chimeras. J Biol Chem 270:10072-8

Tamarappoo BK, Verkman AS (1998) Defective aquaporin-2 trafficking in nephrogenic diabetes insipidus and correction by chemical chaperones. J Clin Invest 101:2257-67

Tang WJ, Stanzel M, Gilman AG (1995) Truncation and alanine-scanning mutants of type I adenylyl cyclase. Biochemistry 34:14563-72

Tang Y, Hu LA, Miller WE, Ringstad N, Hall RA, Pitcher JA, DeCamilli P, Lefkowitz RJ (1999) Identification of the endophilins (SH3p4/p8/p13) as novel binding partners for the β_1-adrenergic receptor. Proc Natl Acad Sci USA 96:12559-64

Tarasova NI, Rice WG, Michejda CJ (1999) Inhibition of G-protein-coupled receptor function by disruption of transmembrane domain interactions. J Biol Chem 274:34911-5

Teller DC, Okada T, Behnke CA, Palczewski K, Stenkamp RE (2001) Advances in determination of a high-resolution three-dimensional structure of rhodopsin, a model of g-protein-coupled receptors (gpcrs). Biochemistry 40:7761-72

Tharapel AT, Anderson KP, Simpson JL, Martens PR, Wilroy Jr RS, Llerena Jr JC and Schwartz CE (1993) Deletion $X_{q26.1-->q28}$ in a proband and her mother: molecular characterization and phenotypic-karyotypic deductions. Am J Hum Genet 52:463-71

The genome international sequencing consortium (2001) Initial sequencing and analysis of the human genome. Nature 409:860-921

Thompson JB, Wade SM, Harrison JK, Salafranca MN, Neubig RR (1998) Cotransfection of second and third intracellular loop fragments inhibit angiotensin AT_{1a} receptor activation of phospholipase C in HEK-293 cells. J Pharmacol Exp Ther 285:216-22

Tiberi M, Caron MG (1994) High agonist-independent activity is a distinguishing feature of the dopamine D_{1B} receptor subtype. J Biol Chem 269:27925-31

Tonacchera M, Chiovato L, Pinchera A, Agretti P, Fiore E, Cetani F, Rocchi R, Viacava P, Miccoli P, Vitti P (1998) Hyperfunctioning thyroid nodules in toxic multinodular goiter share activating thyrotropin receptor mutations with solitary toxic adenoma. J Clin Endocrinol Metab 83:492-8

Tonacchera M, Van Sande J, Cetani F, Swillens S, Schvartz C, Winiszewski P, Portmann L, Dumont JE, Vassart G, Parma J (1996) Functional characteristics of three new germline mutations of the thyrotropin receptor gene causing autosomal dominant toxic thyroid hyperplasia. J Clin Endocrinol Metab 81:547-54

Tribioli C, Droetto S, Bione S, Cesareni G, Torrisi MR, Lotti LV, Lanfrancone L, Toniolo D, Pelicci P (1996) An X chromosome-linked gene encoding a protein with characteristics of a rhoGAP predominantly expressed in hematopoietic cells. Proc Natl Acad Sci USA 93:695-9

Tu JC, Xiao B, Yuan JP, Lanahan AA, Leoffert K, Li M, Linden DJ, Worley PF (1998) Homer binds a novel proline-rich motif and links group 1 metabotropic glutamate receptors with IP₃ receptors. Neuron 21:717-26

Tulman ER, Afonso CL, Lu Z, Zsak L, Kutish GF, Rock DL (2001) Genome of lumpy skin disease virus. J Virol 75:7122-30

Turki J, Pak J, Green SA, Martin RJ, Liggett SB (1995) Genetic polymorphisms of the ß₂-adrenergic receptor in nocturnal and nonnocturnal asthma. Evidence that Gly16 correlates with the nocturnal phenotype. J Clin Invest 95:1635-41

Ullmer C, Schmuck K, Figge A, Lubbert H (1998) Cloning and characterization of MUPP1, a novel PDZ domain protein. FEBS Lett 424:63-8

Ulloa-Aguirre A, Stanislaus D, Arora V, Vaananen J, Brothers S, Janovick JA, Conn PM (1998) The third intracellular loop of the rat gonadotropin-releasing hormone receptor couples the receptor to G_s- and $G_{q/11}$-mediated signal transduction pathways: evidence from loop fragment transfection in GGH3 cells. Endocrinology 139:2472-8

Ulrich CD 2nd, Holtmann M and Miller LJ (1998) Secretin and vasoactive intestinal peptide receptors: members of a unique family of G protein-coupled receptors. Gastroenterology 114:382-97

Unger VM, Hargrave PA, Baldwin JM, Schertler GF (1997) Arrangement of rhodopsin transmembrane a-helices. Nature 389:203-6

Unson CG, Cypess AM, Kim HN, Goldsmith PK, Carruthers CJ, Merrifield RB, Sakmar TP (1995) Characterization of deletion and truncation mutants of the rat glucagon receptor. Seven transmembrane segments are necessary for receptor transport to the plasma membrane and glucagon binding. J Biol Chem 270:27720-7

Vaisse C, Clement K, Guy-Grand B, Froguel P (1998) A frameshift mutation in human MC₄R is associated with a dominant form of obesity. Nat Genet 20:113-4

Valverde P, Healy E, Jackson I, Rees JL, Thody AJ (1995) Variants of the melanocyte-stimulating hormone receptor gene are associated with red hair and fair skin in humans. Nat Genet 11:328-30

van Geel PP, Pinto YM, Voors AA, Buikema H, Oosterga M, Crijns HJ, van Gilst WH (2000) Angiotensin II type 1 receptor A1166C gene polymorphism is associated with an increased response to angiotensin II in human arteries. Hypertension 35:717-21

van Lieburg AF, Verdijk MA, Schoute F, Ligtenberg MJ, van Oost BA, Waldhauser F, Dobner M, Monnens LA, Knoers NV (1995) Clinical phenotype of nephrogenic diabetes insipidus in females heterozygous for a vasopressin type 2 receptor mutation. Hum Genet 96:70-8

Varrault A, Le Nguyen D, McClue S, Harris B, Jouin P, Bockaert J (1994) 5-Hydroxytryptamine 1A receptor synthetic peptides. Mechanisms of adenylyl cyclase inhibition. J Biol Chem 269:16720-5

Venkatesh B, Ning Y, Brenner S (1999) Late changes in spliceosomal introns define clades in vertebrate evolution. Proc Natl Acad Sci USA 96:10267-71

Venter JC et al. (2001) The sequence of the human genome. Science 29:1304-51

Vezza R, Habib A, FitzGerald GA (1999) Differential signaling by the thromboxane receptor isoforms via the novel GTP-binding protein, Gh. J Biol Chem 274:12774-9

Vila-Coro AJ, Rodriguez-Frade JM, Martin De Ana A, Moreno-Ortiz MC, Martinez-A C, Mellado M (1999) The chemokine SDF-1α triggers CXCR4 receptor dimerization and activates the JAK/STAT pathway. FASEB J 13:1699-710

Villar AJ, Pedersen RA (1994) Parental imprinting of the Mas protooncogene in mouse. Nat Genet 8:373-9

Wajnrajch MP, Gertner JM, Harbison MD, Chua SC Jr, Leibel RL (1996) Nonsense mutation in the human growth hormone-releasing hormone receptor causes growth failure analogous to the little (lit) mouse. Nat Genet 12:88-90

Walz T, Hirai T, Murata K, Heymann JB, Mitsuoka K, Fujiyoshi Y, Smith BL, Agre P, Engel A (1997) The three-dimensional structure of aquaporin-1. Nature 387:624-7

Wang Z, Hipkin RW, Ascoli M (1996) Progressive cytoplasmic tail truncations of the lutropin-choriogonadotropin receptor prevent agonist- or phorbol ester-induced phosphorylation, impair agonist- or phorbol ester-induced desensitization, and enhance agonist-induced receptor down-regulation. Mol Endocrinol 10:748-59

Ward DT, Brown EM, Harris HW (1998) Disulfide bonds in the extracellular calcium-polyvalent cation-sensing receptor correlate with dimer formation and its response to divalent cations in vitro. J Biol Chem 273:14476-83

Weinstein LS, Yu S (1999) The role of genomic imprinting of Ga in the pathogenesis of Albright Hereditary Osteodystrophy. Trends Endocrinol Metab 10:81-5

Weitz CJ, Miyake Y, Shinzato K, Montag E, Zrenner E, Went LN, Nathans J (1992) Human tritanopia associated with two amino acid substitutions in the blue-sensitive opsin. Am J Hum Genet 50:498-507

Wenkert D, Schöneberg T, Merendino JJ Jr, Rodriguez Pena MS, Vinitsky R, Goldsmith PK, Wess J, Spiegel AM (1996) Functional characterization of five V_2 vasopressin receptor gene mutations. Mol Cell Endocrinol 124:43-50

Wess J (1998) Molecular basis of receptor/G-protein-coupling selectivity. Pharmacol Ther 80:231-64

West AP Jr, Llamas LL, Snow PM, Benzer S, Bjorkman PJ (2001) Crystal structure of the ectodomain of Methuselah, a Drosophila G protein-coupled receptor associated with extended lifespan. Proc Natl Acad Sci USA 98:3744-9

Westphal RS, Backstrom JR, Sanders-Bush E (1995) Increased basal phosphorylation of the constitutively active serotonin 2C receptor accompanies agonist-mediated desensitization. Mol Pharmacol 48:200-5

White DC, Hata JA, Shah AS, Glower DD, Lefkowitz RJ, Koch WJ (2000) Preservation of myocardial ß-adrenergic receptor signaling delays the development of heart failure after myocardial infarction. Proc Natl Acad Sci USA 97:5428-33

White JH, Wise A, Main MJ, Green A, Fraser NJ, Disney GH, Barnes AA, Emson P, Foord SM, Marshall FH (1998) Heterodimerization is required for the formation of a functional GABA$_B$ receptor. Nature 396:679-82

Wildin RS, Antush MJ, Bennett RL, Schoof JM, Scott CR (1994) Heterogeneous AVPR2 gene mutations in congenital nephrogenic diabetes insipidus. Am J Human Genetics 55:266-77

Wildin RS, Cogdell DE, Valadez V (1998) AVPR2 variants and V2 vasopressin receptor function in nephrogenic diabetes insipidus. Kidney Int 54:1909-22

Williams J, Spurlock G, McGuffin P, Mallet J, Nothen MM, Gill M, Aschauer H, Nylander PO, Macciardi F, Owen MJ (1996) Association between schizophrenia and T102C polymorphism of the 5-hydroxytryptamine type 2a-receptor gene. European Multicentre Association Study of Schizophrenia (EMASS) Group. Lancet 347:1294-6

Wonerow P, Schöneberg T, Schultz G, Gudermann T, Paschke R (1998) Deletions in the third intracellular loop of the thyrotropin receptor. A new mechanism for constitutive activation. J Biol Chem 273:7900-5

Wu V, Yang M, McRoberts JA, Ren J, Seensalu R, Zeng N, Dagrag M, Birnbaumer M, Walsh JH (1997) First intracellular loop of the human cholecystokinin-A receptor is essential for cyclic AMP signaling in transfected HEK-293 cells. J Biol Chem 272:9037-42

Xiao B, Tu JC, Petralia RS, Yuan JP, Doan A, Breder CD, Ruggiero A, Lanahan AA, Wenthold RJ, Worley PF (1998) Homer regulates the association of group 1 metabotropic glutamate receptors with multivalent complexes of homer-related, synaptic proteins. Neuron 21:707-16

Xie J, Murone M, Luoh SM, Ryan A, Gu Q, Zhang C, Bonifas JM, Lam CW, Hynes M, Goddard A, Rosenthal A, Epstein EH Jr, de Sauvage FJ (1998) Activating Smoothened mutations in sporadic basal-cell carcinoma. Nature 391:90-2

Xie Z, Lee SP, O'Dowd BF, George SR (1999) Serotonin 5-HT$_{1B}$ and 5-HT$_{1D}$ receptors form homodimers when expressed alone and heterodimers when co-expressed. FEBS Lett 456:63-7

Yang T, Snider BB, Oprian DD (1997) Synthesis and characterization of a novel retinylamine analog inhibitor of constitutively active rhodopsin mutants found in patients with autosomal dominant retinitis pigmentosa. Proc Natl Acad Sci USA 94:13559-64

Yeagle PL, Alderfer JL, Albert AD (1997) Three-dimensional structure of the cytoplasmic face of the G protein receptor rhodopsin. Biochemistry 36:9649-54

Yeo GS, Farooqi IS, Aminian S, Halsall DJ, Stanhope RG, O'Rahilly S (1998) A frameshift mutation in MC4R associated with dominantly inherited human obesity. Nat Genet 20:111-2

Yuan C, Byeon IJ, Li Y, Tsai MD (1999) Structural analysis of phospholipase A2 from functional perspective. 1. Functionally relevant solution structure and roles of the hydrogen-bonding network. Biochemistry 38:2909-18

Yun J, Schöneberg T, Liu J, Schulz A, Ecelbarger CA, Promeneur D, Nielsen S, Sheng H, Grinberg A, Deng C, Wess J (2000) Generation of a mouse model for X-linked nephrogenic diabetes insipidus. J Clin Invest 106:1361-71

Zawarynski P, Tallerico T, Seeman P, Lee SP, O'Dowd BF, George SR (1998) Dopamine D$_2$ receptor dimers in human and rat brain. FEBS Lett 441:383-6

Zeitlin PL (2000) Pharmacologic restoration of delta F508 CFTR-mediated chloride current. Kidney Int 57:832-7

Zen KH, McKenna E, Bibi E, Hardy D, Kaback HR (1994) Expression of lactose permease in contiguous fragments as a probe for membrane-spanning domains. Biochemistry 33:8198-206

Zeng FY, Soldner A, Schöneberg T, Wess J (1999) Putative disulfide bond in the m3 muscarinic acetylcholine receptor is essential for proper receptor trafficking but not for receptor stability and G protein coupling. J Neurochem 72:2404-14

Zeng FY, Wess J (1999) Identification and molecular characterization of m3 muscarinic receptor dimers. J Biol Chem 274:19487-97

Zhang P, Jobert AS, Couvineau A, Silve C (1998) A homozygous inactivating mutation in the parathyroid hormone/parathyroid hormone-related peptide receptor causing Blomstrand chondrodysplasia. J Clin Endocrinol Metab 83:3365-8

Zhang R, Cai H, Fatima N, Buczko E, Dufau ML (1995) Functional glycosylation sites of the rat luteinizing hormone receptor required for ligand binding. J Biol Chem 270:21722-8

Zhang Z, Sun S, Quinn SJ, Brown EM, Bai M (2000) The extracellular calcium-sensing receptor dimerizes through multiple types of intermolecular interactions. J Biol Chem 276:5316-22

Zhou AT, Assil I, Abou-Samra AB (2000) Role of asparagine-linked oligosaccharides in the function of the rat PTH/PTHrP receptor. Biochemistry 39:6514-20

Zhu SZ, Wang SZ, Hu J, el-Fakahany EE (1994) An arginine residue conserved in most G protein-coupled receptors is essential for the function of the m1 muscarinic receptor. Mol Pharmacol 45:517-23

Zhu X, Gudermann T, Birnbaumer M, Birnbaumer L (1993) A luteinizing hormone receptor with a severely truncated cytoplasmic tail (LHR-ct628) desensitizes to the same degree as the full-length receptor. J Biol Chem 268:1723-8

Zhu X, Wess J (1998) Truncated V_2 vasopressin receptors as negative regulators of wild-type V_2 receptor function. Biochemistry 37:15773-84

Zielenski J (2000) Genotype and phenotype in cystic fibrosis. Respiration 67:117-33

Zitzer H, Honck HH, Bachner D, Richter D, Kreienkamp HJ (1999) Somatostatin receptor interacting protein defines a novel family of multidomain proteins present in human and rodent brain. J Biol Chem 274:32997-3001

Zitzmann M, Nieschlag E (2000) Hormone substitution in male hypogonadism. Mol Cell Endocrinol 161:73-88

Zuscik MJ, Porter JE, Gaivin R, Perez DM (1998) Identification of a conserved switch residue responsible for selective constitutive activation of the ß$_2$-adrenergic receptor. J Biol Chem 273:3401-7